100 COMPREHENSIVE SOLUTIONS IN ALGEBRA

A COMPILATION OF 100 STEP BY STEP PROGRESSIVE ALGEBRA SOLUTIONS

By : Maria Cristina Aquino Santander

    MCAS EDUCATIONAL

INTRODUCTION

This book is a compilation of 100 solutions covering major topics in Advanced Algebra. Such topics include Word Problems involving age, ratio, mixture, investment, number, geometry, work, motion; applications of rational equations, solving systems of linear equations, quadratic equations, quadratic function, exponential function, logarithmic function, arithmetic and geometric sequences, finding equation of a line, circle, parabola, ellipse, variation problems, imaginary numbers.

This compilation is designed to help secondary students in learning the techniques involved in solving algebra word problems. This compilation presents comprehensive and step by step progressive solutions. This compilation can also be a supplementary reference. This material serves as one of the basic reference guide of MCAS Educational Training Center (Manila, Philippines).

I. Solving Age Problems

One of the most popular applications of linear equations is solving age problems. In this segment I present three word problems of this sort with their complete solution. Hope you will enjoy this. .

Problem Number One:

A mother's age is five years greater than twice her son's age as of the present. Fifteen years ago, the mother's age was six times her son's age. What are the present ages of the mother and the son?

Solution:

Representation:

Their present ages:

Let X = Son's age

$2X + 5$ = Mother's age

Their ages 15 years ago:

$X - 15$ = Son's age

$2X + 5 - 15 = 2X - 10$ = Mother's age

Working Equation:

$2X - 10 = 6(X - 15)$

$2X - 10 = 6X - 90$

$90 - 10 = 6X - 2X$

$80 = 4X$

$80/4 = 4X/4$

$20 = X$

$X = 20$

2X + 5 = 2(20) + 5 = 45

The mother is 45 years old and the son is 20 years old as of the present.

Problem Number Two:

A grandpa's age is five times his grandson's age. Ten years from now the grandpa's age will be three times his grandson's age. Find their present ages.

Solution:

Representation :

Their present ages:

Let X = grandson's age

5X = grandpa's age

Their ages ten years from now

X + 10 = grandson's age

5X + 10 = grandpa's age

Working equation:

5X + 10 = 3 (X + 10)

5X + 10 = 3X + 30

5X − 3X = 30 − 10

$2X = 20$

$2X/2 = 20/2$

$X = 10$

$5X = 50$

Grandpa's age is 50 while his grandson's age is 10 years as of the present.

Problem Number Three:

Fred is three years older than his sister Mary who is eleven years old as of the present. In how many years will Mary's age be six-sevenths of Fred's present age?

Solution:

Representation:

Let X = the number of years Mary's age will be 6/7 of her brother's age

Their ages after X years

$11 + X$ = Mary's age in X years

$14 + X$ = Fred's age in X years

Working Equation:

$11 + X = 6/7 \ (14 + X)$

7(11 + X) = (6/7(14 + X)) 7

77 + 7X = 6(14) + 6X

77 + 7X = 84 + 6X

7X – 6X = 84 – 77

X = 7 years.

Mary's age will be 6/7 of her brother's age in seven years.

Solving Age Problems Part Two

This selection is a sequel to my previous segment "Solving Age Problems." In this segment, I present additional four problems with their solution. I hope you will enjoy this.

Problem Number One:

Andrew is nine year older than three times his son's age. In fifteen years, the age of Andrew will be six years more than twice his son's age. Find their present ages.

Solution:

Representation:

Let X = Son's age

3X + 9 = Andrew's age

Their ages in 15 years or fifteen years from now

X + 15 = Son's age

3X + 9 + 15 = 3X + 24 = Andrew's age

Working Equation:

3X + 24 = 2(X + 15) + 6

3X + 24 = 2X + 30 + 6

3X – 2X = 36 – 24

X = 12 = Son's age

3X + 9 = 3(12) + 9 = 45 = Andrew's age

Problem Number Two:

Linda is two thirds as old as Francis. In five years, the sum of their ages will be fifty years. Find their present ages.

Solution:

Representation:

Let X = Francis age

2/3 X = Linda's age

Their ages in five years

X + 5 = Francis age

2/3 X + 5 = Linda's age

Working Equation:

$X + 5 + 2/3 X + 5 = 50$

$X + 2/3 X + 10 = 50$

$X + 2/3 X = 50 - 10$

$(X + 2/3 X = 40) 3$

$3X + 2X = 120$

$5X = 120$

$5X/5 = 120/5$

$X = 24 =$ Francis age

$2/3 X = 16 =$ Linda's age

Problem Number Three:

In eight years, Alvin will be three times as old as he was eight years ago. How old is Alvin now?

Solution:

Representation:

Let $X =$ Alvin's age

$X + 8 =$ Alvin's age in 8 years

$X - 8 +$ Alvin's age eight years ago

Working Equation:

$X + 8 = 3(X - 8)$

$X + 8 = 3X - 24$

$X - 3X = -24 - 8$

$-2X = -32$

$-2X/-2 = -32/-2$

$X = 16$

Alvin is sixteen years old now.

Problem Number Four:

A man who is 42 years old has a daughter who is 12 years old. In how many years will the father be twice as old as his daughter?

Solution:

Man's present age = 42

Daughter's present age = 12

Let X = in how many years will the father's age be equal to twice his daughter's age

Their ages in X years from now

42 + X = Father's age

12 + X = Daughter's age

Working equation:

$42 + X = 2(12 + X)$

$42 + X = 24 + 2X$

$X - 2X = 24 - 42$

$(-X = -18) -1$

$X = 18$

In eighteen years, Father's age will be twice his daughter's age.

II. Solving Word Problems Involving Ratios

Word Problems involving ratios are one of the basic applications of Algebra. In this segment, I presented several problems involving ratios. Hope you will find these ones challenging.

Sample Problem Number One:

The ratio of Edward's money to Gary's money was 3:2. After spending two hundred pesos each at the internet café, the ratio became 2:1. How much did each have before going to the Internet café?

Solution:

Let $3X$ == Edward's original amount of money

$2X$ == Gary's original amount of money

Working Equation:

Since each spent two hundred pesos at the Internet café :

(3X − 200)/ (2X − 200) = 2

3X − 200 = 2(2X − 200)

3X − 200 = 4X - 400

400 - 200 = 4X - 3X

X = 200

3(200) = 600 pesos ===Edward's original money

2(200) = 400 pesos === Gary's original money

Check: Ratio after each one spent 200 pesos in the Internet cafe

600 − 200 = 400

400 - 200 = 200

400/200 = 2/1

Sample Problem Number Two:

Originally, in Miss Angela Perez special science class, the ratio of boys to girls is 3:4.

From this class, five boys and five girls were eliminated and transferred to regular class. The ratio then of boys to girls becomes 2:3. How many students are there in the original class of Miss Perez?

Let 3X === Original number of boys

4X === Original number of girls

Working Equation:

According to the problem, five boys and five girls transferred to regular class and the ratio becomes 2:3;

$(3X – 5)/(4X – 5) = 2/3$

After Cross multiplying we have,

$3(3X – 5) = 2(4X – 5)$

$9X - 15 = 8X - 10$

$9X – 8X = 15 - 10$

$X = 5$

$3(5) = 15$ == Original number of boys

$4(5) = 20$ == Original number of girls

Original class size = 15 + 20 = 35

Sample Problem Number Three:

Out of 200 people who joined the pleasure trip to SAMKARA Resort , some were teachers and the rest were students. If 40 students will be added to this group, the ratio of the number of students who joined to the number of teachers present will become 5:1. How many teachers joined this pleasure trip?

Let X = Number of teachers who joined

200 - X = Number of students who joined

Working Equation:

X/ ((200 - X) + 40) = 1/5

5X = 200 - X + 40

5X + X = 240

6X = 240

6X/6 = 240/6

X = 40

There were 40 teachers who joined this pleasure trip.

Sample Problem Number Four:

In a certain private hospital, there are 50 attending physicians. This hospital maintains the ratio of the number of patients to the number of doctors as 20:1. How many more doctors should the hospital hire to reduce this ratio to 10:1 ?

Given:

50 = Number of doctors

20 (50) = Number of patients considering the ratio 20:1

Let N = Additional number of doctors to be hired to reduce ratio to 10:1

Working Equation:

(50 + N) / 1000 = 1/10

10(50 + N) = 1000

500 + 10N = 1000

10N = 1000 - 500

10N = 500

10N/10 = 500/10

N = 50

The hospital needs to hire 50 more doctors to reduce the ratio of the numbers of patients to the numbers of doctors to 10:1.

III.   **Solving Mixture Problems**

One of the most important and interesting applications of Algebraic equations is found in solving mixture problems. I find solving mixture problems challenging but a little bit confusing. In this segment, I presented several mixture problems with their solutions. Hope you will enjoy reading this.

Problem Number One:

How many gallons of a liquid that is 74 percent alcohol must be combined with 5 gallons of another liquid that is 90 percent alcohol to obtain a mixture that is 84 percent alcohol?

Solution:

Let X = the number in gallons of the first liquid

First Liquid:

Number of gallons: X

Percent of Alcohol: 74%

Number of gallons of alcohol: 0.74X

Second Liquid:

Number of gallons: 5 gallons

Percent of Alcohol: 90%

Number of gallons of Alcohol: (.90)(5) = 4.5 gallons

Mixture:

Number of gallons: X + 5

Percent of Alcohol: 84%

Number of gallons of alcohol: 0.84(X + 5)

Working Equation:

The number of gallons of alcohol in the first liquid + The number of gallons of alcohol in the second liquid = The number of gallons of alcohol in the mixture.

0.74 X + 4.5 = 0.84 (X + 5)

0.74X + 4.5 = 0.84X + 4.2

0.74X - 0.84X = 4.2 - 4.5

-0.10 X = -0.3

10X = 30

1/10(10X) = 30 (1/10)

X = 3

Hence the required amount of first liquid is 3 gallons.

Problem Number Two:

A contractor mixed two batches of concrete that were 9.3 percent and 11.3 percent cement to obtain 4,500 lbs of concrete that was 10.8 percent cement. How many pounds of each type of concrete was used ?

Solution:

Let X = pounds of concrete with with 9.3% cement

4,500 – X = pounds of concrete with 11.3 % cement

First Concrete:

Amount in pounds: X

Percent of Cement: 9.3%

Number of pounds cement: 0.093X

Second Concrete:

Amount in pounds: 4,500 – X

Percent of Cement: 11.3%

Number of pound cement: (0.113)(4,500 – X)

Mixture

Amount in pounds: 4,500

Percent of Cement: 10.8%

Amount of cement in pounds: 0.108(4,500) = 486 pounds

Working Equation:

The amount in pounds of cement in concrete one + the amount in pound of cement in concrete two = Amount in pounds of cement in the mixture.

0.093X + 0.113(4,500 − X) = 486

0.093X + 508.5 − 0.113X = 486

0.093X - 0.113X = 486 - 508.5

- 0.02 X = -22.5

2X = 2250

½(2X) = 2250(1/2)

X = 1,125 lbs

4500 - X = 3,375 lbs.

There were 1125 lbs of concrete with 9.3 % cement while 3375 lbs with 11.3% cement used.

Problem Number Three:

British Sterling is 7.5 percent copper by weight. How many grams of silver must be mixed with 159 grams of an alloy that is 10 percent copper in order to make sterling?

Solution:

Let X = Number of grams of silver that must be mixed with the alloy

Alloy:

Number of grams: 150 grams

Percent Copper: 10%

Amount of Copper in grams: (0.10) (150) = 15 grams

Silver:

Number of grams: X

Percent Copper: 0%

Amount of Copper: 0

Sterling:

Number of grams: 150 + X

Percent Copper: 7.5%

Amount of Copper: 0.075 (150 + X)

Working Equations:

Grams of Copper in Alloy = Grams of Copper in Sterling

$15 = 0.075(150 + X)$

$15 = 11.25 + 0.075 X$

$15 - 11.25 = 0.075 X$

$3.75 = 0.075 X$

$375 = 7.5 X$

$(1/7.5) 375 = 7.5 X (1/7.5)$

X = 50 grams

50 grams of silver must be mixed with 150 grams of alloy which is 10% copper in order to make a sterling.

SOURCE:

COLLEGE ALGEBRA By

Rees

Sparks

Rees

Solving Mixture Problems Part Two

This segment is a sequel to the previous segment, "Solving Mixture Problems". In this segment, I present additional four problems with their solution. Hope you will enjoy this.

Problem Number One:

How many gallons of water must be evaporated from 100 gallons of 75 % salt solution to increase the concentration to 90 %?

Solution:

Let X = Amount of water that must be evaporated

Original solution:

Amount of solution: 100 gallons

Amount of salt in the solution: 75 % of 100 gallons

Amount of water in the solution: 25 % of 100 gallons

Resulting solution:

Amount: 100 – X

Amount of salt: 90% of (100-X)

Amount of water 10% 0f (100-X)

Working equation:

Amount of water in the original solution minus the amount of water evaporated =

Amount of water in the resulting solution after evaporation

$.25(100) - X = .10(100 - X)$

$25(100) - 100X = 10(100 - X)$ ====➔ Multiplying by 100

$2500 - 100X = 1000 - 10X$

$-100X + 10X = 1000 - 2500$

$-90X = -1,500$

$90X = 1,500$

$(1/90) 90X = 1,500 (1/90)$

$X = 16.67$ gallons

The amount of water that must be evaporated is about 16.67 gallons

Problem Number Two:

A radiator that holds 16quartz is full of a solution of 30% alcohol. How much of this solution must be drawn off and replaced with pure alcohol in order that the contents of the radiator may be 55% alcohol?

Solution:

Let X = Amount of original solution that must be drawn and replaced with pure alcohol

Original solution:

Amount of solution: 16quartz

Amount of alcohol: 30% of 16 quarts

Amount of water: 55 % of 16 quartz

Resulting solution:

Amount of solution: 16quartz

Amount of alcohol: 55% of 16quartz

Amount of water: 45% of 16quartz

Working equation:

Amount of alcohol in original minus amourt of alcohol in a 30 percent alcohol solution plus

 a pure alcohol equals amount of alcohol in the final alcohol solution.

.30(16) - .30X + X = .55(16)

30(16) - 30X +100X = 55(16)

480 + 70X = 880

70X = 880 – 480

70X/70 = 400/70

X = 5.71 quartz

Problem Number Three:

A 200 millimeter shampoo with 80% cleansing power is to be diluted with water. This is done by drawing out some amount and replacing it with water. Agnes wants a mild shampoo with 60% cleansing power. How much must she draw off and replace?

Solution:

Let X = Amount of solution must be drawn off and replace

Original solution:

Amount of solution: 200 milliliters

Amount of cleansing power: 80% of 200 ml

Amount of water: 20% of 200 ml

Resulting Solution:

Amount of solution: 200 ml

Amount of cleansing power: 60% of 200 ml

Amount of water: 40% of 200 ml

Working equation:

.20(200) - .20X + X = .40 (200)

20 (200) – 20X + 100 X = 40 (200)

4000 + 80X = 8000

80X = 8000 – 4000

80X = 4000

80X/80 = 4000/80

X = 50 milliliters

Problem Number Four:

Lynn, a chemist mixed 40 mL of 8% HCl acid with 60 mL of 12% HCl acid solution. She used a portion of this solution and replaced it with distilled water. If the new solution tested 5.2 % HCl acid, how much of the original solution did she use ?

Let X = Amount of original solution used and replaced with distilled water.

Step One: Find first the percentage HCl in the original mixture

.08(40) + .12(60) = X 100

8(40) + 12(60) = 10000X

320 + 720 = 10,000X

1,040 = 10,000X

10,000X = 1,040

10,000X/10,000 = 1,040 /10,000

X = 10. 4% HCl in the original mixture

Step two:

Amount of original solution: 100mL

Amount of HCl: 10.4% of 100 mL

Amount of Water: 89.6 % of 100 mL

Resulting solution:

Amount: 100 mL

Amount HCl: 5.2 % of 100 mL

Amount of water: 94.8% of 100 mL

Working equation:

.896(100) - .896X + X = .948(100)

896(100) – 896 X + 1000X = 948(100)

89,600 + 104X = 94,800

104X = 94,800 – 89,600

104X = 5,200

104X/104 = 5,200/104

X = 50 milliliters of the original mixture must be drawn and replaced with distilled water.

IV. Solving Investment Problems

One of the most important applications of linear equations is found in solving investment problems. Investment problems use the Simple Interest formula I = Prt, where

P = Principal (Amount invested)

r = rate of simple interest

t = time in years

In this segment, I present several problems with their solutions.

Problem Number One:

A man invested 35,000 pesos partly at 15% and the rest at 18%. The total income from these investments is 5,650 pesos. Find the sum invested at each rate.

Solution:

Let X = Amount invested at 15%

35,000 – X = Amount invested at 18%

Since Interest = Prt therefore

.15X = Interest earned at amount invested at 15%

.18(35,000 – X)= Interest earned at amount invested at 18%

Interest earned at 15% + Interest earned at 18% = 5,650

Working Equation:

.15X + .18(35,000 – X) = 5,650

Multiplying this equation by 100

15X + 18(35,000 – X) = 565,000

15X + 630,000 - 18X = 565,000

-3X = 565,000 – 630,000

(-3X = -65,000) -1/3

X = 21,667.67 ➔ Amount invested at 15%

35,000 – 21,666.67 = 13,3333.33 => Amount Invested at 18%

Problem Number Two:

A church congregation has 20,000 pesos to be invested in a fund , part at 3% and part at 7%. If the investment at 7% earns

500 pesos more per year than the other placement, how much is invested at each rate?

Solution:

Let X = Amount invested at 3%

20,000 − X = Amount invested at 7%

Interest earned at 3% = .03X

Interest earned at 7% = .07(20,000 − X)

Interest at 3% + 500 = Interest at 7%

Working equation:

.03X + 500 = .07(20,000 − X)

Multiplying this equation by 100

3X + 50,000 =7(20,000 −X)

3X + 50,000 = 140,000 − 7X

3X + 7X = 140,000 − 50,000

(10X = 90,000) 1/10

X = 9,000 === Amount invested at 3%

20,000 − 9,000 = 11,000 === Amount invested at 7%

Problem Number Three:

Mr. Alfonso invested a sum of money at 4% simple interest and another sum which is 100,000 more than the first sum at 5%.

The annual income from the two investments is 17,000 pesos. How much was Mr. Alfonso's investment at each rate?

Solution:

Let X = Amount invested at 4%

X + 100,000 = Amount invested at 5%

Interest at 4% = .04X

Interest at 5% = .05 (X + 100,000)

Interest at 4% + Interest at 5% = 17,000

Working equation:

.04X + .05(X + 100,000) = 17,000

Multiplying this equation by 100

4X + 5 (X + 100,000) = 1,700,000

4X + 5X + 500,000 = 1,700,000

9X = 1,700,000 – 500,000

(9X = 1,200,000) 1/9

X = 133,333.33 => Amount invested at 4%

X + 100,000 = 133,333.33 + 100,000 = 233,333.33 ➔Amount invested at 5%

V. SOME INTERESTING MATH PROBLEMS

I was searching through my files finding some good Mathematics materials when I came across these Math Exercises. I chose some interesting problems that are of some topics in Algebra and Probability. I present here the problems with their solutions.

Problem Number One :

An apple, an orange, a banana and a pear are laid out in a straight line . The orange is not at either end and is somewhere to the right of the banana. In how many ways can the fruit be laid out ?

Solution :

The orange (O) must be on the second or third place from the left. The banana (B) must be somewhere to the left of the orange. Hence the placement of the banana and the orange may take any of three forms namely BO_ _ , B_O_, or _ BO_. In each case two ways remain to fill in the open positions with an apple (A) and a pear (P). The total number of ways equals 3*2 or 6. The ways can be listed as follows:

BOAP BPOA PBOA BOPA BAOP ABOP

Problem Number Two :

Near the end of a party , everyone shakes hands with everybody else. A straggler arrives and shakes hands with only those people whom the straggler knows . Altogether sixty-eight handshakes occurred. How many other people at the party did the straggler know ?

Solution :

If all n people at a party shake hands with all others present then n(n-1)/2 handshakes will take place altogether. Hence, the number of handshakes before the straggler's arrival must have been sixty-six because that is the largest plausible value less than sixty-eight. The straggler must have known two other people at the party. Constructing a table of possible values of n(n-1/2 clarifies that sixty-six is the only plausible number.

N	n(n-1)/2
7	21
8	28
9	36
10	45
11	55
12	66
13	78

Problem Number Three:

Note that 1647/8235 = 1/5, start with 1647/8235, and delete one digit from both the numerator and the denominator to create an equivalent fraction. Then delete another pair to create another equivalent fraction.

Solution:

The successive equivalent fractions are 167/835 and 17/85. The author Barry R. Clarke notes that this fraction is the only "sequential digital deletion fraction" with four digits in both the numerator and denominator that includes eight different digits.

Problem Number Four:

The supplement of an angle is 78 degrees less than twice the supplement of the complement of the angle. Find the measure of the angle,

Solution:

Let A = be the measure in degrees of the angle

$180 - A$ = be the supplement of this angle

$90 - A$ = be the complement of the angle

Working equation:

$(180 - A) + 78 = 2(180 - (90 - A))$

$258 - A = 2(90 + A)$

$258 - A = 180 + 2A$

$78 = 3A$

$A = 26$

Here are another five interesting Math problems. Hope you will enjoy it.

Problem Number One:

It costs $4.00 to have a shave at Arthur's Barber Shop and it cost $7.00 to have a haircut . One Sunday 14 people went to the shop and three of them had both a haircut and a shave. If Arthur earned $98.00 that day, how many did he shave?

Solution: Let X = number of people who had a haircut only

Y = number of people who had both a haircut and a shave

Z = number of people who had shave only

Since a haircut costs $7, a shave cost $4 and both $11 we have

$98 = 7X + 11Y + 4Z$ equation (1) But Y = 3 and since $X + Y + Z = 14$, we get $X + Z = 11$ or $X = 11 - Z$ We plug in the values of Y and X in equation (1) $98 = 7(11 - Z) + 11(3) + 4Z$ so $98 = 77 - 7Z + 33 + 4Z$ Then $98 - 77 - 33 = -3Z$, we get $Z = 4$, the number had a shave only But 3 persons who had a haircut also had a shave so the total number of person who had a shave is $4 + 3 = 7$

Problem Number Two:

2 dozens of apples and 4 dozens of mangoes cost 52 dollars. 3 dozens of mangoes and 4 dozens of bananas cost 59 dollars. If apples and bananas cost the same how much does each mango cost?

Solution:

Since apples and bananas cost the same let their cost denoted b X and the cost of mangoes be Y, So we have (1) $2X + 4Y = 52$

(2) $3Y + 4X = 59$

From equation (1) we obtain $X = (52 - 4Y)/2$ and substitute this for X in (2) we get $3Y + 4(52 - 4Y)(1/2) = 59$ therefore $5Y = 45$ or $Y = 9$

A dozen of mangoes cost 9 dollars therefore a piece of mango costs 9/12 = .75 dollars.

Problem Number Three:

Eight teams entered in the first conference of the NBA. If one team plays the other teams once and only once, how many games will be played in the first conference?

Solution:

Let us say Team 1 is one of those who entered. Team 1 will play the 7 other teams once, so there are seven games, at least. Now team 2 already played Team 1, so it will only have 6 games remaining. Team 3 already played Teams 1 and 2, so it will have 5 more games. Team 4 already played Teams 1, 2 and 3 so it will play 4 more games. By the same reasoning, team 5 will have 3, Team 6 will have 2 and Team 7 will meet Team 8 for the last game. Therefore there are: 7 + 6 + 5 + 4 + 3 + 2 + 1 = 28 games in all.

Problem Number Four:

The Incredible Hulk can dig a tunnel 4 days longer than Spiderman can dig one. If they both work together, they can dig a tunnel in 4 4/5 days. How fast can Spiderman dig a tunnel?

Solution: This is a Work Problem

Let X = number of days can Spiderman dig a tunnel

X + 4 = number of hays can Incredible Hulk dig a tunnel

Working equation: (1/X + 1/(X + 4)) 24/5 = 1

(24/5X + 24/(5X + 20) = 1) 5X(5X + 20)

$24(5X + 20) + 24(5X) = 25X^2 + 100X$

$120X + 480 + 120X = 25X^2 + 100X$

$240X + 480 = 25X^2 + 100X$

$(25X^2 - 140X - 480 = 0) \, 1/5 =\langle$

$5X^2 \; 28X - 96 = 0$ By factoring we get $(5X + 12)(X - 8) = 0$

$===\langle X - 8 = 0 =\langle X = 8$ Answer: 8 days

Problem Number Five:

A deadly epidemic has swept the small planet ZNNX - 8. On the first day 10 died, on the second day 20 died and on the third 40 died. If the progression of mortality continues this way, in how many days will the planet population be wiped out if the population of ZNNX -8 has 20,470 inhabitants. This is a problem involving geometric sequences.

$(20,470 = (10) \, 2 \wedge(n-1)) \, 1/10$

$2047 = 2\wedge(n-1)$

$\log 2047 = \log 2\wedge(n-1)$

$(\log 2047 = (n - 1) \log 2) \, 1/\log 2$

$\log 2047 / \log 2 = n - 1$

$3.31/.3 = n - 1$

$11 = n - 1$

$N = 12$ days.

SOURCE:

A MATH QUIZ BOOK "BRAIN BLITZ" BY Ramon Lorenzo

VI. TOPIC: PROBLEM SOLVING INVOLVING RATIONAL EQUATION

FEATURING:

WORK PROBLEM

MOTION PROBLEM

INTRODUCTION: Why a lesson in solving rational equation?

Rational Equations is one of the most important concepts in Algebra. This subject has sought many great applications in many fields of Mathematics.

In Trigonometry, applications involving solving oblique triangle require the use of rational equation. The Law of Lines, the Law of Cosines and Law of Tangents are usually expressed as a rational equation.

In Physics, majority of the application problems involve rational equations. Most of the problems in Mechanics involve skill in manipulation of rational equations.

May it be a problem involving motion or force, it requires skill in handling rational equations.

In Mathematics of Investment, applying the concept of rational equations remains inevitable. Problems in interest and discount require skills in handling rational equations.

In fact, all fields of Mathematics require handling of rational equations. Next to linear equation, I believe firmly that The concept of "Rational Equations" is the most important concept in Algebra.

LESSON PROPER:

Part One:

Problem Involving Work

Problem Number One:

Leonor mows a lawn in 4 hours Alex can mow the same lawn in 5 hours. How long would it take both of them, working together, to mow the lawn.

Solution:

In solving this problem, one consider first how much of the job is done in one hour, two hours and so on.

It takes Leo 4 hours to mow the entire lawn. Thus after one hour, he has done ¼ of the lawn. It takes Alex 5 hours to mow the entire lawn. Thus after one hour he has done 1/5 of the entire lawn.

Representation:

Let t = the number of hours it will take both Leo and Alex

to finish the lawn working together

Working equation:

t (1/4 + 1/5) = 1

$(t/4 + t/5 = 1)\ 20$

$5t + 4t = 20$

$9t/9 = 20/9$

$t = 20/9$ or 2 and 2/9 hours

Answer : 2-2/9

Sample Problem Two:

It takes Ramil 9 hours longer to construct a fence than it takes Ency. If they work together, they can construct the fence in 20 hours. How long would it takes each working along to construct the fence?

Solution:

Representation:

Let t = the amount of time it would take Ency working alone.

$t + 9$ = the amount of time it would take Ramil working alone

Using the same reasoning in the previous problem, we can assume then that Ency has finished 1/t of the fence after one hour and Ramil has finished 1/t+9 of the fence after one hour. Since Ramil and Ency can complete the entire fence in 20 hours, we have:

$20\ (1/t + 1/t+9) = 1$

as our working equation:

20/t + 20/t+9 = 1   LCM: (t) (t + 9)

(t) (t + 9) (20/t + 20/t+9) = 1 (t) (t + 9)

(t + 9)20 + t . 20 = (t) (t + 9)

20 + 180 + 20t = t2 + 9t

40T + 180 = t2 + 9t

0 = t2 – 31t – 180

(t – 36) (t + 5) = 0

t – 36 = 0   t + 5 = 0

t = 36   t = -5

It takes 36 hours for Ency to do the fence alone and it takes 45 hours for Ramil to do the work alone.

Check: 20/36 + 20/45 = 1

5/9 + 4/9 = 1

9/9 = 1

Part Two: MOTION PROBLEM

Sample Problem:

Biker A bikes 15 km.hr faster than Biker B. By the time it takes Biker A to reach 80 km. Biker B has gone 50 km. Find the speed of each biker.

Solution:

Representation:

Let r = The rate in km/hr. of biker B

$r + 15$ = the rate in km/hr of Biker A

Distance Speed Time

A   80 $r + 15$ t

B   50   r   t

Since time = distance/rate

Distance Speed   Time

A   80 $r + 15$ $80/r+15$

B   50 r $50/r$

From the problem, it is given that the time is the same, we can now have our working equation as:

50/t = 80/r+15

r (r+15) = 50/r = 80/r+15 = (r) (r+15)

50 (r+15) = 80R

50r + 750 = 80r

750 = 80r – 50r

750 = 80r – 50r

750/30 = 30r/30

r = 25

Biker B is going at 25 km/hr. and Biker A's speed is 40 km./hr.

Source :

Mastering Intermediate Algebra

Simon L Chua

Benson S. Tan

Roberto G. Degolacion

Ma. Salome B. Aguinaldo

VII .Solving Word Problems Involving Motion

Among the most challenging applications of linear equations are word problems involving motion problems. The fundamental formula used for this problem is:

d = r t

where d represents distance , r represents rate or velocity or speed and t represents time.

When this formula is used , d and r must use the same unit of distance (for example miles and miles per hour) and r and t must use the same unit of time (for example miles per hour and hours). If the formula is solved for r and t , we get two additional forms of the formula :

r = d/t or t = d/r.

Sample Problem Number One:

Suppose that a trip from the dormitory to the lake at 30 miles per hour takes 12 minutes longer than the return trip at 48 miles per hour. How far apart are the dormitory and the lake?

Solution:

Units used must be the same; so we use 12 minutes = 12/60 = 1/5hr.

Working equation:

Since the speeds are given in miles per hour the equation is :

$d/30 = d/48 + 1/5$

Where d is the distance of the trip in miles and each of the fractions represents time in hours, using $t = d/r$ from above.

$30 \, (d/30) = (d/48 + 1/5) \, 30$

$d = 5d/8 + 6$

$8d = 5d + 48$

$8d - 5d = 48$

$3d = 48$

$d = 16$

The distance between the dormitory and the lake is 16 miles.

Check:

The time to the lake at 30 miles per hour is $16/30 = 32/60 = 32$ minutes.

The time from the lake at 48 miles per hour is $16/48 = 1/3 = 20/60 = 20$ minutes,

Which is indeed 12 minutes quicker.

Problem Number Two:

Dominic rode his motorbike 20 minutes to Helen's home and then the two drove in a car 30 minutes to a beach 35 miles from Dominic's home. If the car speed was 10 miles per hour faster than that of a motorbike, how fast did the car travel?

Motorbike car

_____|_____

—

-------------------------------35 miles------------------------------------

Travelling by Motorbike:

Time: 20min/60 = 1/3 hr

Distance: (1/3) r since d = r t

Rate or Velocity: r

Travelling by Car:

Time: 30 minutes = 30/60 = 1/2 hr

Distance: (r + 10) 1 /2

Rate or Velocity: r + 10

Working Equation:

Distance traveled by motorbike + Distance traveled by Car = Total Distance (35miles).

1/3 r + 1/ 2 (r + 10) = 35

6 {1/3 r + 1/ 2 (r + 10) = 35}

2r + 3r + 30 = 210

5r = 210 - 30

(1/5) 5r = 180 (1/5)

r = 36 miles per hour = speed of the motorbike

r + 10 = 46 miles per hour = speed of the car

Check:

(1/3) (36) + (1/2) (46) = 35

12 + 23 = 35

35 = 35

Sample Problem Number Three:

Two students are 350 km apart and begin walking toward one another at constant rates. One travels at 1.6 m/ sec and the other at 1.9 m/ sec. How long will they walk until they meet? How far has each of them gone?

A ---------------------------350km ---------------------------B
_____

_____

Travel made by student A;

Rate: 1.6 m/sec

Distance: d

Time : d/ 1.6 since t = d/r

Travel made by student B:

Rate: 1.9 m/sec

Distance: 350 – d

Time: (350 – d)/1.9

Working Equation: The time covered by each student is equal

Time covered by student A = Time covered by student B

d/ 1.6 = (350 – d)/ 1.9

1.9 d = 1.6 (350 – d)

1.9 d = 560 – 1.6 d

1.9d + 1.6 d = 560

(1/3.5) 3.5 d = 560 (1/3.5)

d = 160 meters = distance covered by A

350 – 160 = 190 = distance covered by B

How long will they walk until they meet? 160/1.6 = 10 seconds

190/1.9 = 10 seconds

Check: (1.9) (10) + (!.6) (10) = 350

190 + 160 = 350

Sample Problem Number Four:

A bicycle club left the campus to ride to a park 24 miles away for an outing. Lyn left from the same place by car with picnic supplies 1 and a half later, traveled at a speed four times as fast and arrived at the park at the same time as the cyclists. How fast did Lyn drive?

Travel made by the cyclists

Distance: 24 miles

Time: $t + 1.5$

Rate: $24/(t + 1.5)$ since $r = d/t$

Travel made by Lyn

Distance: 24 miles

Time: t

Rate: $24/t$

Working Equation:

Rate of Lyn = 4 * rate of the cyclists

$(t * t + 1.5)(24/t = 4 * 24/(t + 1.5))$

$24(t + 1.5) = 96 t$

$24t + 36 = 96t$

$96t - 24t = 36$

$(1/72)\, 72t = 36\, (1/72)$

$t = 1/2$ hour covered by Lyn

1 and 1/2 + 1/2 = 2 hours = covered by the cyclists

Rate of Lyn: 24/0.5 = 48 miles/hour

Check:

$d = rt$

$24 = 48\, (1/2)$

$24 = 24$

SOURCE:

COLLEGE ALGEBRA By

Rees

Sparks

Rees

VIII. Solving Work Problems

Algebraic equations have sought wide applications in solving various word problems. One of its common applications are done in solving work problems. In this article, I have selected four work problems presenting with their solutions.

Problem Number One:

A farmer can plow a field in 5 days using a tractor. A hired man can plow the same field in 7 days using a smaller tractor. How many days will it take for both of them to plow the same field together?

Solution:

Let X = number of days it will take for both the farmer and the hired man to plow the field together.

1/5 = the part plowed in one day by the farmer

1/7 = the part plowed in one day by the hired man

Working equartion: (1/5 + 1/7) X = 1 1 stands for one complete job

. (X/5 + X/7 = 1) 35

7X + 5X = 35

(12X = 35)1/12

X = 35/12 or 2 and 11/12 days

It will take 2 and 11/12 days for both of them to plow the field together.

Problem Number Two:

If in the previous problem, the hired man worked one day with the smaller machine and joined by the farmer with the larger one, how many more days it will take both of them to finish plowing?

Solution: Since the hired man plowed one-seventh of the field in a day, six-sevenths (6/7) remained unplowed.

Let X = number of days it will take the two to finish the job

X/5 = the part plowed by the farmer

X/7 = the part plowed by the hired man

We now have the working equation, X/5 + X/7 = 6/7

(X/5 + X/7 = 6/7) 35

7X + 5X = 30

(12X = 30) 1/12 X = 2 and ½

The farmer and the hired man have still to work for 2 and ½ days to finish plowing the field.

Problem Number Three:

Arthur can dig a tunnel 5 days longer than Michael can dig one. If they both work together, they can dig a tunnel in 8 days. How fast can Michael dig a tunnel?

Let X = number of days it will take Michael to dig one tunnel

X + 5 = number of days it will take Arthur to dig a tunnel

Working equation: (1/X + 1/(X+5)) 8 = 1

(8/X + 8/(X+5) = 1) (X) (X + 5)

$8(X + 5) + 8X = X^2 + 5X$

$8X + 40 + 8X = X^2 + 5X$

$16X + 40 = X^2 + 5X$

$X^2 = 11X - 40 = 0$ Using quadratic formula to solve for X

$X = (11 + SQRT(121 + 4(40)))$ div 2 $X = (11 + 16.76)/2 = 13.88$ days It will take Michael 13.88 days to dig a tunnel.

Problem Number Four:

A swimming pool can be filled in 6 hours and requires 9 hours to drain. If the drain was accidentally left open for 6 hours while the pool was being filled, how long did filling the pool require?

$1/6$ = part of the pool filled after one hour

$1/9$ = part of the pool drained after an hour

Working equation: $(1/6 - 1/9) 6 + X/6 = 1$

After six hours 1/3 of the pool was filled: $(1/6 - 1/9) 6 = 1/3$

2/3 of the pool is still to be filled after 6 hours

$(X/6 = 2/3)6 =$ $X = 4$ hours $6 + 4 = 10$ hours It takes 10 hours for the pool to be filled.

SOURCE:

COLLEGE ALGEBRA By   Rees Sparks Rees

IX. SOLVING WORD PROBLEMS INVOLVING SYSTEM OF EQUATIONS

Many problems that we encounter in applications involve more than one variable. When the relations between the variables involve several equations, then we must solve the equations of the resulting system simultaneously.

In this segment I present three interesting word problems that involve solving system of equations, with their corresponding solution. I hope you will find this hub enjoyable.

Problem No. One:

The sum of the digits of a three-digit number is 13. If the tens and hundreds digits are interchanged, the new number is 90 less than the original, and if the units and hundreds digits are interchanged, the resulting number is 99 less than the original. Find the original number.

Solution to Problem Number One:

Representation:

Let X = the unit digit

Y = the tens digit

Z = the hundreds digit

$x + 10y + 100z$ = the original number

Since the problem involves three variables, we must derive three working equations.

Working equation Number One:

Based from the sum of the digits is 13:

$X + y + z = 13$

Working equation Number Two:

Based from the tens and hundreds digits interchanged resulting to new number

90 less than the original

$X + 10z + 100y = x + 10y + 100z - 90$

Simplify the equation we get

$-90z + 90y = -90$

$-z + y = -1$ = equation two

Working equation Number Three:

Based from the units and hundreds digits are interchanged, resulting to new number 99 less than the original.

$Z + 10y + 100x = x + 10y + 100z - 99$

Simplify this equation we get:

$-99x + 99x = -99$

$-z + x = -1$ equation 3

Our system of equation now consists of

$X + y + z = 13$ eq. 1

$-z + y = -1$ eq. 2

$-z + x = -1$ eq. 3

Solving equation 1 and equation 2 using elimination method by addition to eliminate z

$X + y + z = 13$

$\underline{-z + y = -1}$

$X + 2y = 12$ equation 4

Solving eqn 1 and eqn 3 using elimination method by addition to eliminate z:

$X + y + z = 13$

$\underline{-z + x = -1}$

$Y + 2x = 12$ - eqn. 5

We now have two new equations

$X + 2y = 12$ - eqn 4

$2x + y = 12$ - eqn. 5

Multiply eqn. 4 by two

$(x + 2y = 12)\,2$ --- $2x + 4y = 24$

Solving eqn 4 and 5 using elimination method by subtraction to eliminate x

$2x + 4y = 24$

$\underline{2x + y = 12}$

$2x + 4y = 24$

$\underline{-2x + y = -12}$

$(1/3)\ 3y = 12)1/3)$

$y = 4$

Substituting y = 4 to either eqn 4 or 5 to solve for x

$2x + 4 = 12$

$\underline{-y = -4}$

$(1/2)\ 2x = 8\ (1/2)$

$X = 4$

Substituting x = 4, y = 4 to corresponding variables in equation 1

$4 + 4 + z = 13$

$8 + z = 13$

$z = 13 - 8$

$z = 5$

The original number is 544.

Problem Number two:

Twenty five coins, whose value is $2.75 are made up of nickel, dime and quarters. If the nickel were dimes, and dimes were quarters, and the quarters nickels, the total value would be $3.75. How many coins each type are there:

Solution:

Representation

Let n = number of nickels

d = number of dimes

q = number of quarters

Working equation number one:

Since there are 25 coins mentioned, the first working equation will be

n + d + q = 25 - eqn. 1

Working equation number two:

Is based on the value of money of $2.75

$(.05)n + (.10)d + (.25)q = 2.75$

Multiplying the whole equation by 100

5n + 10d + 25q = 275 ------- eqn. 2

Working equation number three:

Is based on interchanging nickels and dimes, dimes and quarters, quarters and nickels giving new value of $.375.

$(.10)n + (.25)d + (.05)q = 3.75$

10n + 25d + 5q = 375 …. Eqn. 3

Our system of equation now consists of

n + d + q = 25 --…eqn. 1

5n + 10d + 25q = 275 … eqn. 2

$10n + 25d + 5q = 375$ ….. eqn. 3

Solving eqn 1 and eqn 2:

Multiplying eqn 1 by 5

$(n + d + q = 25)5$

Solving eqn 1 and 2 using elimination method by subtraction

$5n + 5d + 5q = 125$

$\underline{-5n - 10d - 25q = -275}$

$- 5d - 20q = 150$

Multiply $-5d - 20q = -150$ by -1

$5d + 20q = 150$ --- dividing the whole eqn by 5 we get $d + 4q = 30$

Let this be equation 4

Solving eqn 2 and 3

Multiplying eqn 2 by 2

$(5n + 10d + 25q = 275) 2$

$10\backslash n + 20d + 50q = 550$

Solving eqn 2 and 3 using elimination method by subtraction

$10n + 20d + 50 = 5550$

$\underline{-10n - 25d - 5q = -375}$

$- 5d + 45q = 175$

Dividing the whole equation

-5d + 45q = 175 x 5

We get -d + 9q = 35 --- let this be equation 5

Adding equation 4 and 5

d + 4q = 30 …..eqn. 4

-d + 9q = 35 … eqn. 5

(1/13) 13q = 65 (`1/13)

q = 5

Substitute q = 5 in either eqn 4 0r 5

d + 4(5) = 30

d + 20 = 30

d = 20 - 30 = 10

d = 10

Substituting 1 = 5 , d = 1- in egn. 1

n + 10 + 5 = 25

n + 15 = 25 …. n = 25 - 15 = 10

n = 10

There are 5 quarters, 10 dimes and 10 nickels.

Check: (5) (.25) + (10) (.10) + (10) (.05) = 2.75

1.25 + 1.00 + .50 = 2.75

2.75 = 2.75

Problem Number Three:

If A, B and C work together on a job it will take 1-1/3 hours. If only A and B work, it would take 1-5/7 hours, but if B and C work, it would take 2-2/5 hours. How long would it take each man, working done, to complete the job.

Solution:

Representation:

Let x = number of hours it will take workers A to complete a job.

Let y = number of hours it will take worker B to complete a job.

Let z = number of hours it will take worker C to complete a job.

Working equation one:

(1/x + 1/y + 1/z) (4/3) = 1

(4/3x + 4/3y + 4/3z = 1) 3 xyz

4yz + 4 x z + 4 x y = 3 xyx = eqn. 1

Working equation number twoL1/x + 1/y) 12/7 = 1

(12/7x + 12/7y = 1) 7 x y

12y + 12z = 7xy ---- eqn. 2

Working equation number three:

(1/y + 1/z) 12/5 = 1

(12/5y + 12/5z =1) 5 yz

12y + 12z = 5yz ..... eqn. 3

Our system of equation now consist of:

$4yz + 4xz + 4xy = 3xyz$ .....eqn. 1

$12y + 12x = 7xy$ .... eqn. 2

$12z + 12y = 5yz$ .... eqn. 3

Using equation 3 to solve for z

$12z - 5zy = -12y$

$z(12-5y)/12-5y = -12y/12-5y$

$z = -12y/ 12-5y$

Substitute this value of z in eqn 1.

$4y(-12y/12-5y) + 4x(-12y/12-5y) + 4 \times y = 3xy(-12y/12-5y)$

$(-48y2 /12-5y - 48xy /12-5y + 4x y = -36y2x/ 12-5y)(12-5y)$

$-48y2 - 48 \times y + 4 \times y (12-5y) = 36y2 \times$

$-48y2 - 48 \times y + 48 \times y - 20 xy2 = -36 y2 \times$

$-48 y2 = -36y2x + 20 y2 \times$

$(-48y2 = -16 y2 x) -1/16 y2)$

$3 = x$

$x = 3$

Substituting x = 3 to eqn. 2 to solve for y

$12 y + 12(3) = (7)(3)y$

$12y + 36 = 21y$

$36 = 21y - 12y$

$36/9 = 9y/9$ $y = 4$

Substituting y = 4, in eqn. 3 to solve for z

$12z + 12(y) = 5(4)z$

$12z + 48 = 20z$

$48 = 20z - 12z$

$1/8 (48) = 8z (1/8)$

$Z = 6$

It will take 3 hours for worker A, 4 hours for worker B and 6 hours for worker C to complete the job.

Check: $(1/3 + 1/4 + 1/6) 4/3 = 1$

$4/9 + 4/12 + 4/18 = 1$

$16/36 + 12/36 + 8/36 = 1$

$36/36 = 1$

SOURCE:

MODERN ALGEBRA AND TRIGONOMETRY BY VANCE

X. Solving Geometry Problems Involving System of Equations In Two Variables

Algebra has also found wide applications in Geometry. Among its common problems are problems involving perimeters and angles. In this hub I included five sample problems involving perimeter and angles.

Sample Problem Number One:

A rectangular garden has a perimeter of 100 meters. The length is four times its width. Find the dimensions of the garden.

Solution:

Let X = width

Let Y = length

P = 2L + 2w

Equation one: 2X + 2Y = 100

Equation Two: Y = 4X

Substitute equation two in equation one :

2X + 2 (4X) = 100

2X + 8X = 100

10X = 100

(1/10) 10X = 100 (1/10)

X = 10 ➔ width

4X = 4 (10) = 40 === length

Sample Problem Two:

The perimeter of a rectangular picture frame is 80 centimeter. Two times the width is equal to the length increased by five. Find its dimensions.

Solution:

Let X = width

Let Y = length

P = 2L + 2W

Equation one: 2X + 2Y = 80

Equation two: 2X = Y + 5

We use elimination method by subtraction to solve for the unknown variables.

2X + 2Y = 80

(-)2X (+) -Y = (-) 5

3Y = 75

(1/3) 3Y = 75 (1/3)

Y = 25 ➔ length

To solve for width substitute in equation two

2X = 25 + 5

2X = 30

X = 15 ==→ width

Sample Problem Three:

The perimeter of a rectangular flower garden is 120 meters. The length is ten meters greater than its width. Find its dimensions.

Solution:

Let X = width

Let Y = length

P = 2L +2W

Equation one: 2X + 2Y = 120

Equation two: Y = X + 10

Substitute equation two in equation one :

2X + 2 (X + 10) = 120

2X + 2X + 20 = 120

4X = 120 – 20

4X = 100

(1/4) 4X = 100 (1/4)

X = 25 → width

Y = 25 + 10 = 35 =→ length

Sample Problem Number Four:

The sum of the two non right angles in a right triangle is of course 90 degrees. If twice the first is 40 degrees more than three times the second. Find the measurement of the angles of the right triangle.

Solution:

Let X = first angle

Let Y = second angle

Equation one: X + Y = 90

Equation two: 2X = 3Y + 40

From equation one we derive Y = 90 – X the substitute this in equation two.

2X = 3(90-X) + 40

2X = 270 – 3X + 40

2X + 3X = 270 + 40

5X = 310

(1/5) 5X = 310 (1/5)

X = 62 degrees =➔ first angle

Y = 90 – 62 = 28 =➔ second angle

Sample Problem Number Five:

Two angles are supplementary. The bigger angle is thrice as large as the smaller angle. Find the measurement of the angles.

Solution:

Let X = smaller angle

Let Y = bigger angle

The two angles are supplementary it means that their sum is equal to 180 degrees.

Equation one : X + Y = 180

Equation two : Y = 3X

Substitute equation two in equation one:

X +3X = 180

4X = 180

(1/4) 4X = 180 (1/4)

X = 45 ➔ smaller angle

Y = 3(45) = 135 ==➔. Bigger angle

XI. Solving Number Problems Involving System of Equations In Two Variables

One of the most common applications of equations are number problems. This hub presents five number problems involving system of equations in two variables.

Problem Number One:

The sum of two numbers is 70. The smaller number increased by twenty equals the bigger number. What are these two numbers?

Solution:

Let X = smaller number

Let Y = bigger number

Equation one : X + Y = 70

Equation Two: X + 20 = Y

Substitute equation 2 in equation one:

X + X + 20 = 70

2X + 20 = 70

2X = 70 – 20

2X = 50

(1/2) 2X = 50 (1/2)

X = 25

Y = 25 + 20 = 45

The two numbers are 25 and 45

Problem Number Two:

The sum of two numbers is 100. The bigger number is four times as large as the smaller number. Find the two numbers.

Solution:

Let X = smaller number

Let Y = bigger number

Equation one : X + Y = 100

Equation two: Y = 4X

Substituting Equation two in equation one:

X + 4X = 100

5X = 100

(1/5) 5X = 100 (1/5)

X = 20

Y = 4(20) = 80

The two numbers are 20 and 80

Problem Number Three:

A number is thrice as large as the other number. The smaller number is 25 less than half the larger number. Find these numbers.

Solution:

Let X = smaller number

Let Y = bigger number

Equation one: Y = 3X

Equation Two: X = y/2 – 25

Substituting equation one in equation two:

$X = 3X/2 - 25$

$2 (X = 3X/2 - 25) 2$

$2X = 3X - 50$

$2X - 3X = -50$

$-1 (-X = - 50)$

$X = 50$

$Y = 3 (50) = 150$

The two numbers are 50 and 150

Sample Problem Number Four:

A certain number is ten times as large as the other number. This number is 50 greater than eight times the smaller number. Find these two numbers.

Solution:

Let X = smaller number

Let Y = the larger number

Equation one: $Y = 10X$

Equation Two: $Y = 8X + 50$

Substituting Equation one in equation two:

$10X = 8X + 50$

$10X - 8X = 50$

2X + 50

(1/2) 2X = 50 (1/2)

X = 25

Y = 10 (25) = 250

The two numbers are 25 and 250

Sample Problem Number Five:

The sum of two numbers is 70. The smaller number increased by 90 is equal to thrice the bigger number. Find these two numbers.

Solution:

Let X = smaller number

Let Y = bigger number

Equation one: X + Y = 70

Equation Two : X + 90 = 3Y

We will solve this using the elimination method by subtraction

X + Y = 70

(-) X - (+) 3Y = (+) -90

(1/4) 4Y = 160 (1/4)

Y = 40

X = 70 – 40 = 30

The two numbers we are looking for are 30 and 40.

XII. Solving Work Problems Involving System of Linear Equations

In this segment I presented several challenging work problems involving system of linear equations complete with solution. I hope this selection will be useful to you and you will enjoy viewing this.

Sample Problem Number One:

Two machines are used with production of toys. 1000 toys will be produced in a day if machine A operates for 4 hours and machine B operates for 3 hours or if machine A operates for 6 hours and machine B operates for 2 hours. How long would it take each machine to produce 1000 toys.

Solution:

Let X = Number of hours it will take machine A to produce 1000 toys.

Let Y = Number of hours it will take machine B to operate 1000 toys.

$4/X + 3/Y = 1$ equation one

$6/X + 2/Y = 1$ equation two

Multiply equation one by 3 and equation two by -2 to eliminate X and solve for Y:

$12/X + 9/Y = 3$

$-12/X - 4/Y = -2$

$5/Y = 1$ ➔ $Y = 5$ hours

To solve for X you may use substitution method : Using equation one substitute $Y = 5$

$4/X + 3/5 = 1$

$4/X = 1 - 3/5$

$4/X = 2/5$

$2X = 20$

$X = 10$ hours

It will take 10 hours for machine A and 5 hours for machine B to complete 1000 toys each.

Sample Problem Number Two:

Crew #1 and #2 can build a house in 45 days, #2 and #3 for the same job in 36 days, #1 and #3 for the same job in 60 days. If crew #1, #2,#3 work together, how many days can they do the same job ?

Solution:

Let A = number of hours it will take crew #1 to build the house

Let B= number of hours it will take crew#2 to build the house

Let C = number of hours it will take crew #3 to build the house

Let D = number of hours it will take crew #1,#2,#3 to build the house together

45/A + 45/B = 1 equation one

36/B + 36/C = 1 equation two

60/A + 60/C = 1 equation three

D/A + D/B + D/C = 1 equation four

Use equation one and two to eliminate B

(45/A + 45/B = 1) 36

(36/B + 36/C = 1) -45

1620/A + 1620/B = 36

-1620/B - 1620/C = -45

1620/A - 1620/C = -9 let this be equation 5

Now use equation 3 and equation 5 to eliminate either A or C

1620/A - 1620/C = -9

(60/A + 60/C = 1) 27

1620/A - 1620/C = -9

1620/A + 1620/C = 27

3240/A = 18

18A = 3240

A = 180 days

To solve for B substitute A = 180 in equation one:

45/180 + 45/B = 1

¼ + 45/B = 1

45/B = 1 – ¼

45/B = ¾

3B = 180

B = 60 days

To solve for C substitute A = 180 in equation three

60/180 + 60/ C = 1

1/3 + 60/C = 1

60/C = 1 – 1/3

60/C = 2/3

180 = 2C

C = 90 days

To solve for D substitute A = 180 B = 60 C = 90 in equation four

(D/180 + D/60 + D/90 = 1) 180

D + 3D + 2D = 180

6D = 180

D = 30 days

It will take 30 days for the three crews to build the house.

Sample Problem Number Three:

A tank is supplied by two pipes A and B and drained by pipe C. If the tank is full and A and C are opened the tank can be emptied in 10 hours. If B and C are opened the tank can be emptied in 13 hour and 20 minutes. If the tank is empty A and B are opened the tank can be filled in 4 and 4/9 hours. If all pipes are opened , how long it will be filled up.

Solution:

Let W = Number of hours it will take pipe A to fill the tank

Let X = Number of hours it will take pipe B to fill the tank

Let Y = Number of hours it will take pipe C to drain the tank

Let Z = If all pipes are open, the number of hours it will take to fill the tank

$10/W - 10/Y = 1$ equation one

$(40/3)/X - (40/3)/Y = 1$ equation two

$(40/9)/W + (40/9)/X = 1$ equation three

$Z/W + Z/X - Z/Y = 1$ equation four

We will solve this using elimination method first. Using equation one and equation three to eliminate W:

$(10/W - 10/Y = 1)$ multiply by 40/9

$((40/9)/W + (40/9)/X = 1)$ multiply by -10

$(400/9)/W - (400/9)/Y = 40/9$

$-(400/9)/w - (400/9)/X = -10$

$(-400/9)/Y - (400/9)/X = -50/9$ let this be equation five

Now use equation two and five to eliminate either X or Y

$((40/3)/X - (40/3)/Y = 1)$ multiply by 400/9

$((-400/9)/X - (400/9)/Y = -50/9)$ multiply by 40/3

$(16000/27)/X - (16000/27)/Y = 400/9$

$(-16000/27)/X - (16000/27)/Y = -2000/27$

$(-32000/27)/Y = 400/9 - 2000/27$

$(-32000/27)/Y = (1200 - 2000)/27$

$(-32000/27)/Y = -800/27$

$800Y = 32000$

$Y = 40$ hours

To solve for W substitute $Y = 40$ in eqn one

$10/W - 10/40 = 1$

$10/W = 1 + ¼$

$10/W = 5/4$

$5W = 40$

$W = 8$ hours

To solve for X substitute $Y = 40$ in eqn two:

$(40/3)/X - (40/3)/40 = 1$

$40/3X = 1 + 1/3$

$40/3X = 4/3$

$4x = 40$

$X = 10$ hours

To solve for Z substitute $W = 8$, $X = 10$, $Y = 40$ in equation four

$(Z/8 + Z/10 - Z/40 = 1)$ multiply by 40

$5Z + 4Z - Z = 40$

$8Z = 40$

Z = 5 hours

If pipe A, B and C are open, it will take five hours to fill the tank.

This problem set is taken from Ateneo de Manila's High School assignment.

XIII. Solving Word Problems Involving Exponential Function

Exponential function is one of the most important concepts in Algebra. In this segment I present several problems involving exponential function with their solution. I hope you will find this selection useful and interesting.

Problem Number One:

Suppose a culture of 500 bacteria are put in a petri dish and the culture double very hour. How many bacteria will be left after ten hours?

Solution:

Formula to be used:

$P_n = P(1 + r)^n$

Where P=original population

r = rate of growth

n=period

P_n = Expected population after growth

Given: The rate of increase is 100% or 1.

$P_{10} = P(1 + 1)^{10}$

$P_{10} = (500)(2)^{10}$

$P_{10} = 500(1024)$

$P_{10} = 512,000$

There will be 512,000 bacteria after ten hours.

Problem Number Two:

A certain radioactive substance decays half of itself everyday. Initially there are 20 grams. How much substance will be left after ten days?

Solution:

Given: The rate of decay is i/2 or 0.5

$P_{10} = P(1 - 0.5)^{10}$

$P_{10} = 20(0.5)^{10}$

$P_{10} = 20(.000976)$

$P_{10} = 0,01953$ grams

Problem Number Three:

A certain town has a population of 50,000. Its rate increases 8% every six months. Find the population after four years?

Solution:

Given:

$P = 50,000$

$R = .08$

$n = (4)(2) = 8$

$P = 50,000(1.08)^8$

$P = 50,000(1.85)$

P92,546

There will be about 92,546 people after four years.

Problem Number Four:

Paolo deposits 20,000 pesos in a bank that pays 3% compound interest annually. How much money will he have after twelve year without withdrawal /

$A = P(1 + r/m)^{mt}$

A= Total amount after t years

P= Principal amount

r= interest arte

m= number of time the amount is compounded a year

P=20,000

r= .03

m=1

t=12

$P = 20,000(1 + .03)^{12}$

P = 20,000 (1.426)

P =28,515.28

Paolo's money in the bank will be about 28, 515.218 after twelve years.

Problem Number Five:

The half-life of a radioactive substance is 12 hours and there are 100 grams initially. Determine the amount of substance remaining after one week.

The half-life of a radioactive substance is the amount of the time it takes for half of the substance to decay.

The exponential decay formula is:

$A = A_o (1/2)^{t/k}$

Number of hours in one week = (24)(7) = 168 hours

$A = 100(1/2)^{168/12}$

$A = (100)(1/2)^{14}$

$A = 100(.000061035)$

$A = 0.00610$ grams

SOURCE:

ADVANCED ALGEBRA, TRIGONOMETRY AND STATISTICS BY

Orines

Esparrago

Reyes

XIV. Solving Word Problems Involving Logarithms

The use of logarithms has found many applications in various practical transactions of everyday life.. The use of logarithms has been very helpful in dealing with some mathematical problems encountered from time to time. In this hub, I present several word problems which involve the use of logarithm with their complete solution.

Problem Number One: Food Source

The coding model for a coffee served in a mug is TF = TR + (TO - TR) e^-30t.

Given that the original temperature of the coffee is 155 degrees farenheit and the room temperature is 75 degrees farenheit, determine after how many minutes the coffee will be 110 degrees farenheit?

Given:

TF (F nal Temperature) = 110 degrees farenheit

TO (Original Temperature) = 155 degrees farenheit

TR (Room Temperature) = 75 degrees farenheit

e =2.72

Formula:

$TF = TR + (TO - TR) e^{-.30t}$

Substitute the given in the formula:

$110 = 75 + (155 - 75) e^{-.30t}$

$110 - 75 = (80) e^{-.30t}$

$35/80 = e^{-.30t}$

$0.4375 = e^{-.30t}$

$\log 0.4375 = \log e^{-.30t}$

$-0.359 = -.30t \log (2.72))$

$-0.359 = -.30t (0.4346)$

$-0.359/0.4346 = -.30t$

$-0.826/-0.30 = t$

$t = 2.75$ minutes

The coffee will be 110 degrees farenheit after 2.75 minutes.

Problem Number Two: Sales

The model for predicting the sales S of a new brand of sweat shirt is :

S = 50,000 - 50,000 e ^-rt, where r is the rate of growth of sales. Determine the growth rate of sales to the nearest tenth of a percent if 4000 sweat shirts were sold in the fast two years.

Given:

e = 2.72

S = 4000

t = 2 years

r = ?

Formula:

S = 50,000 – 50,000 e^-rt

Substitute the given in the formula

4000 = 50,000 - 50,000 e^-2r

4000- 50,000 = -50,000e^-2r

-46,000/50,000 = - 50,000e^-2r/50,000

0.92 = e^-2r

Log 0.92 = log e^-2r

-0.0362 = -2r log log 2.72

-0.0362 = -2r 0.4346

-0.0362/0.4346 = -2r

-0.0833/-2 = r

r = 0..04 or 4 percent

Growth rate of sales is at 4 percent.

Problem Three: Finance

Solve the equation $1.05^n = 1.08^7$, which shows how many years it would take for money invested at 5 percent compounded anually to equal the value of money invested at 8 percent compounded annually for seven years.

$1.05^n = 1.08^7$

$\log 1.05^n = \log 1.08^7$

$n \log 1.05 = 7 \log 1.08$

$n \log 1.05 / \log 1.05 = 7 \log 1.08 / \log 1.05$

n = 7(0.0334)/0.021)

n = 11.13

It will take 11 years.

XV. Solving Word Problems Involving Quadratic Equation

One of the most important topics in Algebra is solving quadratic equations. It has found wide applications in many areas of Mathematics. This hub presents several problems involving quadratic equations with their solution. Hope you will enjoy this.

Sample Problem One:

The sum of two numbers is 18 and the sum of their square is 170. Find the numbers.

Solution:

Let X = the first number

$18 - X$ = the second number

X^2 = square of the first number

$(18-X)^2$ = square of the second number

Working Equation:

$X^2 + (18-X)^2 = 170$

$X^2 + 324 - 36X + X^2 = 170$

$2X^2 - 36X + 324 - 170 = 0$

$(2X^2 - 36X + 154 = 0)1/2$

$X^2 - 18X + 77 = 0$

Factoring $X^2 - 18X + 77 = 0$

$(X - 7)(X - 11) = 0$

$X - 7 = 0$ therefore $X = 7$

$X - 11 = 0$ therefore $X = 7$

The two number are 7 and 11.

Problem Number Two:

One leg of a right triangle is 7 cm shorter than the other leg. Its area is 30cm^2. Find its perimeter.

Solution :

Let X = length of one leg

$X + 7$ = length of the other leg

Area of a triangle = (bh)/2

Working Equation:

X (X+7)/2 = 30

(X(X+7)/2 = 30) 2

X^2 + 7X = 60

X^2 + 7X - 60 = 0

Factoring X^2 + 7X -60

(X -5) (X + 12) = 0

X – 5 = 0 therefore X = 5

Leg1 = 5

Leg2 = 5 + 7 = 12

Finding the hypotenuse:

Hypotenuse = SQRT(5^2 + 12^2) = SQRT(25 + 144) = SQRT (169) =13

Perimeter = 5+12+13 = 30 centimeters.

Problem Number Three:

One side of a triangle is one dm. less than its diagonal. If the perimeter of the rectangle is 62, find the area of the rectangle.

Solution :

Length of the rectangle = d – 1

To represent width use phytagorean theorem and consider diagonal as hypotenuse:

$d^2 = w^2 + (d - 1)^2$

Rearranging the equation : $w^2 = d^2 - (d-1)^2$

$w^2 = d^2 - d^2 + 2d - 1$

$w^2 = 2d - 1$

$w = SQRT(2d - 1)$

Working Equation : Based from Perimeter = 2l + 2W

$2(d-1) + 2\ SQRT(2d - 1) = 62$

$2d - 2 + 2\ SQRT(2d-1) = 62$

$(2SQRT(2d - 1) = 64 - 2d)\ ½$

$SQRT(2d - 1) = 32 - d$

$(SQRT(2d - 1))^2 = (32 - d)^2$

$2d - 1 = 1024 - 64d + d^2$

$d^2 - 66d + 1025 = 0$

Factoring gives $(d - 41)(d - 25) = 0$

$d - 25 = 0$

$d = 25$

Length = $d - 1 = 25 - 1 = 24$

Width = SQRT(2d-1) = SQRT(50- 1) = SQRT49 = 7

Area of the rectangle = (24)(7) = 168 dm^2

XVI. Solving Word Problems Involving Quadratic Function

Some of the most important functions in applications are the quadratic functions. The quadratic functions are one of the simplest type of polynomial function, aside from the linear functions. Applications include satellite dishes, projectile paths, economics, geometry, ecology, weather, biology and many others.

A function f is a quadratic function if $f(x) = ax^2 + bx + c$ where a, b, and c are real numbers and not equal to 0.. The graph of a quadratic function is a parabola. The most basic aid in graphing a parabola is knowing whether a > 0 (the graph opens upward) or a < 0 (the graph opens downward). The two simplest quadratic functions are $f(x) = x^2$ and $g(x) = -x^2$.

In this selection, I feature two interesting problems with their complete solution.

Problem Number One:

Jack Pott dives off the high diving board. His distance from the surface of the water

varies quadratically with the number of seconds that have passed since he left the

board.

(a) His distance at time 1, 2 and 3 seconds since he left the board are 24, 18, 2

meters above the water respectively. Write a particular quadratic equation

expressing the distance in terms of time.

(b) How high is the diving board ?

(c) When does he hit the water ?

Solution:

(a) In finding the quadratic equation three points can be derived from the given :

(1, 24) ; (2, 18) ; (3, 2);

by letting the time in seconds as x-coordinates and the distance above the

water as y-coordinates.

The quadratic equation has a standard form $ax^2 + bx + c = 0$.

Using the given points above, we must solve for a, b, and c.

From the point (1, 24) we can derive an equation by substituting 1 to x's and

24 to y, The equation formed is : $a + b + c = 24$ let this be equation #1.

From point (2, 18) we can derive this equation: $4a + 2b + c = 18$ let this be

equation #2.

From point (3, 2) we can derive this equation : $9a + 3b + c = 2$ let this be

Equation #3.

We will now solve for a, b, and c using elimination method. by subtraction.

Using equation 1 and equation 2 to eliminate c;

$a + b + c = 24$

$-4a - 2b - c = -18$

$-3a - b = 6$ let this be equation 4

Using equation 2 and equation 3 to eliminate c

$4a + 2b + c = 18$

$-9a - 3b + c = -2$

$-5a - b = 16$ let this be equation 5

Solving for a and b using equation 4 and equation 5

$-3a - b = 6$

$+5a + b = -16$

$2a = -10$

$a = -5$

substituting $a = -5$ in equation 4

$-3(-5) - b = 6$

$15 - b = 6$

$b = 15 - 6$

$b = 9$

substituting $a = -5$ $b = 9$ in equation 1 to solve for c

$-5 + 9 + c = 24$

$c = 24 - 4$

$c = 20$

Therefore the quadratic equation we are looking for is $-5x^2 + 9x + 20 = 0$.

(b) How high is the diving board?

In order to solve for the height of the diving board we must solve for the vertex of the parabola $y = -5x^2 + 9x + 20$. This parabola opens downward since a is negative.

Actually, the vertex is the maximum point of the parabola. The height of the diving board is the y-coordinate of the vertex. Solve first for h or x-coordinate of the vertex. The formula to solve for h is -b/2a. Therefore h = -9/(2) (-5) = 9/10.

Substitute 9/10 to every x in the given quadratic function $y = -5x^2 + 9x + 20$.

$y = -5 (9/10)^2 + 9 (9/10) + 20$

$y = -81/20 + 81/10 + 20$

$y = 481/20 = 24.05$ meters.

(c) When does he hit the water?

y = 0, when the diver hits the water so in order to solve for this we must let

$-5x^2 + 9x + 20 = 0$ and solve for x using quadratic formula.

The quadratic formula is given by:

$X = (-b +- SQRT(b\textasciicircum2-4ac))/2a$

X = (-9 + - SQRT(81 – (4)(-5)(20))) /2(-5)

X = (-9 +- SQRT(481))/-10

X = (-9 – 21.93) /-10

X = -30.3/-10

X = 3.03 seconds

Problem Number Two:

A satellite dish in storage has parabolic cross sections and is resting on its

vertex.. A point on the rim is 4 feet high and is 6 feet horizontally from the

vertex. How high is a point which is 3 feet horizontally from the vertex?

Solution:

Plotting this parabola on a rectangular coordinate plane with vertex at (0, 0), we can derive other two points based from the given . Two points on the rim are

(-6, 4) and (6, 4).

Now let us derive the equation of the parabola using points (0, 0) ; (-6, 4), (6, 4).

C = 0

Using point (-6, 4) and (6, 4)

$4 = a(-6)^2 + (-6)b + c$

$4 = 36a - 6b + c$

$4 = 36a + 6b + c$

========================

$8 = 72a$

$a = 1/9$

Solving for b;

$4 = 36(1/9) + 6b$

$4 = 4 + 6b$

$B = 0$

The equation of the parabola is $y = 1/9 \, x^2$

$Y = 1/9 \, (3)^2$

$Y = 1/9(9) = 1$

The point which is 3 feet horizontally from the vertex is 1 foot high.

This problem set is taken from Saint Scholastica's College High School assignment.

Solving Word Problems Involving Quadratic Functions Part Two

This selection is a sequel to the hub "Solving Word Problems Involving Quadratic Functions. In this segment I present three additional application problems. I hope you will enjoy this hub and benefit much from it.

Sample Problem Number One:

The sum of two numbers is 48. Find the maximum product and the two numbers.

Solution:

Let n = The first number

48 – n = the second number

f(n) = product of the two numbers

f(n) = n (48 – n) = 48n – n^2

f(n) = 48n - n ^2

Find the vertex of the parabola:

a = -1 (coefficient of n^2

b = 48 (coefficient of n)

h = -b/2a =è this is the abscissa of the vertex of the parabola

The parabola opens downward and the vertex is the maximum point since a is a negative value.

Find h : h = -48/2(-1) = 24

k = f(24) = 48(24) - 24^2 =1152 - 576 = 576 this is the ordinate of the vertex.

Thus the maximum product is 576. The product of the two numbers n and 48-n must be equal to the maximum product 576. That is:

n(48 – n) = 576

48n - n^2 = 576

n^2 -48n + 576 = 0

(n -24) ^2 = 0

n = 24

If n = 24 then 48-n = 24. Therefore the two numbers whose sum is 48 and whose product is a maximum are 24 and 24.

Sample Problem Number Two:

Find the dimension and the maximum area of a rectangle if its perimeter is 36.

Solution:

Given:

Let l = length

Perimeter P = 36

P = 2l + 2w

36 = 2l + 2w

w = (36 -2l)/2

A = l*w

A = ((36-2l)/2) l

f(A) = 36l/2 - 2l^2/2

f(A) = 18l - l^2

a = -1 b = 18

h = -b/2a = -18/2(-1) = 9

k = f(9) = 18(9) - 9^2 = 162 - 81 = 81 è this is the ordinate of the parabola.

Vertex is at maximum therefore 81 is the maximum area.

Solving for l:

((36 – 2l)/2)l = 81 Multiplying this equation by two

$(36 - 2l)\, l = 162$

$36l - 2l^2 = 162$

$2l^2 - 36l + 162 = 0$ Dividing by two

$l^2 - 18l + 81 = 0$

$(l - 9)^2 = 0$

$l = 9$

$w = (36 - 2l)/2 = (36 - 18)/2 = 18/2 = 9$

The dimensions of the rectangle should be length = 9 and width = 9 to be able to get the maximum area.

Sample Problem Number Three:

The sum of two numbers is 36. Find the numbers whose sum of the squares is a minimum.

Solution:

Let n = first number

36 − n = second number

$F(S) = n^2 + (36 - n)^2$

$F(S) = n^2 + 1{,}296 - 72n + n^2$

$F(S) = 2n^2 - 72n + 1{,}296$ Dividing this equation by two

$F(S) = n^2 - 36n + 648$

$a = 1$ $b = -36$

$h = -(-36)/2 = 18$

$k = f(18) = 18^2 - 36(18) + 648$

$k = 324 - 648 + 648 = 324$ This is the ordinate of the vertex

The vertex of the parabola is at minimum since a is a positive value.

324 is the minimum sum of the squares

$324 = n^2 - 36n + 648$

$n^2 - 36n + 648 - 324 = 0$

$n^2 - 36n + 324 = 0$

$(n - 18)^2 = 0$

$n - 18 = 0 \rightarrow n = 18$

The first number is 18 and the other number is 18. The two numbers which will give a minimum value for the sum of the squares are 18 and 18.

SOURCE:

ADVANCED ALGEBRA, TRIGONOMETRY AND STATISTICS

By :

Orines

Esparrago

Reyes

XVII. Finding The Equation of A Line

One of the most important topics in Algebra is finding the equation of the line. In this selection, I present five problems with their solution.

Problem Number One:

Find an equation of the line passing through (-5, -1) and (3, 3).

Solution:

The point-slope form of the equation of the line through the point (X1, Y1) and slope m is :

Y – Y1 = m (X - X1).

Let us first find the slope m of the line using the formula for finding the slope given two points.

M = Y2 – Y1/X2 – X1

Designate (-5, -1) as (X_1, Y_1) and (3, 3) as (X_2, Y_2).

m = (3 – (-1))/ (3 – (-5)) = (3 +1) /(3+ 5) = 4/8 = ½

Then you may use either of the two points to substitute in the equation :

$Y - Y_1 = m (X - X_1)$.

Using the point (-5, -1) we got

Y – (-1) = ½(X – (-5))

2[Y+ 1 = ½(X + 5)]

2Y + 2 = X + 5

X - 2Y +3 = 0 is the equation of the line we are looking for.

Problem Number Two:

Find the equation of the line which is parallel to the line 3X + 6Y = 7 and passing through (3, -5).

Solution:

Parallel lines have the same slope. We have to solve for the slope of the line we are looking for and substitute it in the point-slope form equation of the line. To solve for the slope we have to convert the equation 3X + 6Y= 7 into slope-intercept form of the line which is Y = m X + b where m is the slope.

1/6[6Y= -3X+7]

Y =-1/2X+7/6

The coefficient of X is the slope m=-1/2, .-1/2 is also the slope of the line we are solving since they are parallel lines. Then substitute m = -1/2 and the point (3, -5)in the point-slope form equation:

2[Y – (-5)= - ½ (X – 3)]

2Y + 10 =X -3

X – 2Y -13 = 0 is the equation of the line we are looking for.

Problem Number Three:

Find an equation of the line passing through (4, 5) which is perpendicular to the line 7X + 6Y = -3.

The slope of a line perpendicular to a certain line is the negative reciprocal of the slope of that certain line perpendicular to the given line . To solve for the slope of 7X + 6Y= -3 we rearrange this equation to slope-intercept form or Y = m X + b.

7X+6Y=-3

[6Y=-7X -3]

Y= -7/6 X- ½

The slope of the line we are solving for is6/7which is negative reciprocal of -7/6. Reciprocal of 7/6 and opposite in sign. Then substitute m = 6/7and (4, 5) to the point-slope form of the line:

$7[Y - 5 = 6/7 (X - 4)]$

$7Y - 35 = 6X - 24$

$6X - 7Y + 11 = 0$ is the equation of the line we are looking for.

Problem Number Four:

Find the equation of the line with the same X-intercept as the line $2X - 9Y = 14$ and parallel to the line $X - Y = 21$,

Solution:

We have to solve for the X-intercept of the line $2X - 9Y = 14$. To solve for the X-intercept let $Y = 0$.

$2X - 9(0) = 14$

$(2X = 14) \, 1/2$

$X = 7$ is the X-intercept.

Since X-intercept = 7, we consider (7, 0) is appoint in the line we are solving. The line is parallel to $X - Y = 21$ so has the same slope as this line. Convert $X - Y = 21$ into slope-intercept form :

$(- Y = -X + 21) -1$

$Y = X + 21$

$m = 1$

We use $m = 1$ and point (7,0) to solve for the line. Substituting this value into point-slope form we get:

Y = X - 7

X – Y – 7 = 0 is the equation of the line we are looking for.

Problem Number Five:

Find the equation of the line with X-intercept as 7 and Y-intercept as -3.

Solution:

(7, 0) and (0,-3) are two points of this line. WE solve first for m.

m=-3 /-7=3/7. Then we use either of the two points to substitute in the point-slope form : Let us use (7, 0):

[y = 3/7(X – 7)] 7

7Y = 3X -21

3X – 7Y -21= 0is the equation of the line we are looking for.

XVIII. Finding The Equation of A Circle

A circle is a set of all points in a plane equidistant from a fixed point which is the center of the circle. The distance from the center to a point on the circle is called the radius.

Standard equation for a circle with center at the origin : $X^2 + Y^2 = r^2$

Standard equation for a circle with center at (h, k) : $(X - h)^2 + (Y - k)^2 = r^2$

General equation of the circle : $X^2 + Y^2 + DX + EY + F = 0$.

Sample Problems Involving Circle:

Problem Number One:

Find the equation of a circle which passes through points (10, 2) ; (3, 9); (-2. 10).

Solution : Consider the general equation of a circle as $X^2 + Y^2 + DX + EY + F = 0$.

Using the given three points we derive our equation:

From the point (10, 2) we get the equation $100 + 4 + 10D + 2E + F = 0$ or

$104 + 10D + 2E + F = 0$ by substituting 10 to X's and 2 to Y's. This is our equation one.

From the point (3, 9) we get $9 + 81 + 3D + 9E + F = 0$ or $90 + 3D + 9E + F = 0$.

From the point (-2.10) we get $4 + 100 - 2D + 10E + F = 0$ or $104 - 2D + 10E + F = 0$.

We now have a system of three equations:

$104 + 10D + 2E + F = 0$ eqn one

$90 + 3D + 9E + F = 0$ eqn two

$104 = 2D + 10E + F = 0$ eqn three

We now use elimination method to find the value of D, E and F.

Using equation one and eqn two to eliminate F, we do this by subtracting equation two from eqn one :

$104 + 10D + 2E + F = 0$

$-(90 + 2D + 9E + F = 0)$

We get $14 + 7D - 7E = 0$ let this be eqn four.

Using Equation 2 and eqn 3 to eliminate F we subtract eqn 3 from eqn 2:

$90 + 3D + 9E + F = 0$

$- (104D - 2D + 10E + F = 0)$

We get $-14 + 5D - E = 0$ let this be equation five

Using eqn four and eqn five we can solve for D and E.

$14 + 7D - 7E = 0$ eqn four

$-14 + 5D - E = 0$ eqn five

Multiple equation five by -7 in order to eliminate E

$(-14 + 5D - E = 0) * (-7) == 98 = 35D + 7E = 0$

Adding eqn 4 and eqn 5

$14 + 7D - 7E = 0$

$98 - 35D + 7E = 0$

$112 - 28D = 0$

$(1/28)\, 112 = 28D\, (1/28)$

$D = 4$

Substitute $D = 4$ in equation four ;

$14 + 7(4) - 7E = 0$

$14 + 28 - 7E = 0$

$42 = 7E$

$E = 6$

Substitute $D = 4$, $E = 6$ in eqn 1 to solve for F

$104 + 10(4) + 2(6) + F = 0$

$104 + 40 + 12 + F = 0$

$156 + F = 0$

$F = -156$

The general equation of the circle we are solving is :

$X^2 + Y^2 + 4X + 6Y - 156 = 0.$

We get this general equation by substituting D = 4, E = 6 and F = -156 to the general equation of the circle $X^2 + Y^2 + DX + EY + F = 0$.

To solve for the standard equation of the circle we will use completing the square:

$X^2 + 4X + \underline{} + Y^2 + 6Y + \underline{} = 156$

To complete the square divide the coefficient of the middle term by two then square it.

$X^2 + 4X + 4 + Y^2 + 6Y + 9 = 156 + 4 + 9$.

The standard equation of the circle we are looking for is:

$(X + 2)^2 + (Y + 3)^2 = 169$

The circle has its center at (-2, -3) and has a radius equal to 13.

Problem Number Two:

Find the equation of a circle whose diameter has its endpoints at A(-3, 5) and B(1, 3).

To find the center of the circle find the midpoint of the diameter using midpoint formula.

Here is the midpoint formula:

$X = (X_1 + X_2)/2$ is used in finding the X-coordinate of the midpoint.

$Y = (Y_1 + Y_2)/2$ is used in finding the Y-coordinate of the midpoint.

Using the endpoints (-3, 5) and (1, 3)

$X = (-3 + 1)/2 = -1$ $Y = (5 + 3)/2 = 4$

Therefore the midpoint of the diameter which is also the center of the circle we are looking for is (-1, 4)

To find the radius of the circle find the distance between the center and one endpoint using distance formula. The distance formula is:

$D = SQRT((Y_2 - Y_1)^2 + (X_2 - X_1)^2)$.

Using one endpoint (1, 3) and the center of the circle (-1, 4), let us find the distance using the distance formula:

$D = SQRT((3-4)^2 + (1 + 1)^2) => SQRT(5)$

Therefore the standard equation of the circle we are looking for is :

$(X + 1)^2 + (Y - 4)^2 = 5$

XIX. Finding the Equation of a Parabola

Basic Definition

A parabola is a set of all points in a plane that are equidistant from a fixed point and a fixed line. The fixed point is called the "focus" and the fixed line is called the "directrix". The focus may not be on the directrix.

The line through the focus and perpendicular to the directrix is called the "axis of symmetry" or just the axis of the parabola. The line segment through the focus and perpendicular to the axis of symmetry which is cut by the parabola is called the focal chord or "latus rectum" and its length is the focal width. The point of intersection of the axis of symmetry and the parabola is called the vertex. It then follows by the definition that the vertex is equidistant from the focus and the directrix.

Standard Forms of the Equations of a Parabola

An equation of the parabola with its vertex at (h, k) and focus at (h + p, k) is :

$(y - k)^2 = 4p(x - h)$.

The axis of symmetry of this parabola is parallel to the X-axis and the graph opens to the right if $p > 0$ and to the left if $p < 0$. The equation of the axis of symmetry is $y = k$

With directrix $x = h - p$.

An equation of the parabola with the vertex at (h, k) and focus at (h, k + p) is:

$(x - h)^2 = 4p(y - k)$.

The axis of symmetry of this parabola is parallel to the Y-axis and the graph opens up if p > 0 and down if p < 0. The equation of the axis of symmetry is x = h with directrix

y = k – p.

If the vertex(h, k) is at the origin then h = k = 0, and the equation has the following forms :

Y ^ 2 = 4px

X ^ 2 = 4py.

The focal width of the parabola is 4p. The distance between vertex and focus is I p I.

Sample Problem Number One:

Find the equation and focal width of the parabola with vertex at (5, 1) and focus at (7, 1).

Solution:

In this problem, the vertex and focus are of the same distance from the X-axis ; therefore the line y = 1 is the axis of symmetry . Hence the form of the equation is (y – k)^2 = 4p (x – h) with h = 5 and k = 1. Since the focus is two units to the right of the vertex , p = 2

4p = 4(2) = 8

Consequently, the equation of the parabola is:

$(y - 1)^2 = 8(x - 5)$ and the focal width is absolute value of 8 = 8.

Sample Problem Number Two:

Find the equation of the parabola given vertex at (5, 0) and y = -8 as directrix.

The axis of symmetry is parallel to the Y –axis. The equation for the directrix is y = k – p. Substituting k = 0 and y = -8 to the equation:

-8 = 0 – p

-8 = -p

8 = p or p = 8

Therefore the equation of the parabola we are looking for is:

$(x - 5)^2 = 8y$

Sample Problem Number Three:

Find the equation of the parabola given directrix y = 3 and focus at (5, -1).

The axis of symmetry of this parabola is parallel to the Y-axis. The parabola opens downward. The vertical distance from the directrix to the vertex is equal to p which is equal to the vertical distance from the vertex to the focus. Therefore vertical distance from the directrix to the focus is 2p. Get the directed distance from the directrix to the focus.

Absolute value of (-1 – 3) = 4

2p = 4

p = 2 thus 4p = 8.

Therefore the vertex is at (5, -1 + 2) or at (5, 1).

The equation of the parabola we are looking for is: $(x - 5)^2 = 8(y - 1)$.

Sample Problem Number Four:

Find the equation of the parabola with vertex at (-2, 3) axis parallel to the X-axis and passing through (4, 9).

The form of the equation must be: $(y - 3)^2 = 4p(x + 2)$.

Substituting x = 4 and y = 9 gives 36 = 4p (6) and so 4p = 6. The equation thus is :

$(y - 3)^2 = 6(x + 2)$.

Sample Problem Number Five:

Find the equation of the parabola given vertex at (5, 2) and ends of the focal chord at (3, 6) and (3, -2).

The axis of symmetry is parallel to the X-axis. The parabola opens to the left .

4p = length of the focal chord.

Get the vertical or directed distance between the two ends of the focal chord.

Absolute value of (-2-6) = 8

4p = 8

Thus the equation of the parabola we are looking for is :

$(y – 2)^2 = -8 (x – 5)$.

SOURCE:

COLLEGE ALGEBRA By Rees Sparks Rees

XX. Finding the Equation of an Ellipse

An ellipse is the set of all points in a plane such that the sum of the distances of each point from two fixed points is the same, The fixed points are called the "foci" and the line through the foci is called the axis of symmetry. The ratio c/a is called the eccentricity of the ellipse. The point(h, k) that is midway between the foci is called the center of the ellipse. The intersection points of ellipse and the line through the foci are known as the vertices. The line segment joining the vertices is the major axis . The part of the line through the center perpendicular to the major axis and intercepted by the ellipse is the minor axis.

Standard Forms of the Equation of an Ellipse

(1) $(X-h)^2/a^2 + (Y-k)^2/b^2 = 1$

Is an equation of the ellipse with center at (h, k), vertices at (h+- a, k) ,foci at

(h +- c, k), major axis parallel to the X-axis and semi axes of length a and b.

(2) $(Y-k)^2/a^2 + (X-h)^2/b^2 = 1$

Is an equation of ellipse with center at (h, k), vertices at (h, k +- a), foci at

(h, k +- c), major axis parallel to the Y-axis and semi axes of length a and b.

If the center is at the origin or (0,0), the equations above have the form :

$X^2/a^2 + Y^2/b^2 = 1$ and

$Y^2/a^2 + X^2/b^2 = 1$

In each case $a^2 = b^2 + c^2$ and thus $a > b$ and $a > c$.

Also for every ellipse a is the distance from center to vertex and c is the distance from center to focus.

Problem Number One:

Find the equation of the ellipse with center at (-1. 1) and a focus at (3, 1) and a vertex at (4, 1).

Solution:

We can write the equation of an ellipse if we know the center, the semi axes and which of the two standard forms to use. With the given data, the center, focus and vertex are on the line parallel to the X-axis therefore the major axis of the ell pse is parallel to the X-axis. We will use the standard equation (1). Since a is the distance between the center and the vertex, we solve for a by just simply getting the directed distance between the X-coordinates of the center and the vertex.

$a = 4 - (-1)$ ==> $a = 4 + 1$ è. $a = 5$

Also, since c is the distance between the center and the focus we solve for c by simply getting the directed distance between the X-coordinates of the center and the focus.

$c = 3 - (-1)$ $c = 3 + 1$ $c = 4$

We now solve for b using the formula $a^2 = b^2 + c^2$

b = SQRT(a^2 − c^2)

b = SQRT(25 -16) = 3

Since now we know the values of a and b we just simply substitute these values to standard equation (1) of an ellipse. Thus

((X + 1)^2)/25 + ((Y − 1)^2)/9 = 1

is the equation of the ellipse we are solving.

Problem Number Two:

Find the equation of the ellipse whose ends of the minor axis are at (-4, 3) and (6,3)

and a vertex at (1, 10).

Solution:

The midpoint of the minor xix is the center of the ellipse.

X = (X1 + X2)/2 = (- 4+ 6)/2 = 2/2 = 1

Therefore (1, 3) is the center of the ellipse whose equation we are solving for.

Since the minor axis is parallel to the X-axis, the major axis of the ellipse we are solving is parallel to the Y-axis.

In order to get a , get the directed distance between the vertex (1,10) and the center (1, 3). Just get the directed distance between the Y-coordinates of the center and vertex.

a = 10 − 3 = 7.

To get b, get the directed distance between the X-coordinates (1, 3) and one endpoint of the minor axis (6, 3);

b = 6 − 1 = 5.

Now we have value for a and b and we have known the center we can now find the standard equation of the ellipse we are solving by simply substituting the designated values to standard equation number (2).

$(Y - 3)^2/49 + (X - 1)^2/25 = 1$

Problem Number Three:

Find the equation of the set of all points such that the sum of the distances of each point from (2, 5) and (2, -3) is 14.

Solution:

By the definition of the ellipse the given points are the foci, 2a is 14 and a = 7. The center is the midpoint of the line joining the foci namely (2, 1). Since 2c = 5 − (-3) = 8

We have c = 4. Thus $b^2 = 7^2 - 4^2 = 49 - 16 = 33$ and the equation we are looking for is :

$(Y - 1)^2/49 + (X - 2)^2/33 = 1$

Problem Number Four:

Put the following equation in standard form and find the center, vertices and foci.

$9X^2 + 4Y^2 - 36X - 8Y + 4 = 0$

Group and Factor:

$(9X^2 - 36X) + (4Y^2 - 8Y) = -4$

$9(X^2 - 4X) + 4(Y^2 - 2Y) = -4$

Complete the square:

$9(X^2 - 4X + 4) + 4(Y^2 - 2Y + 1) = -4 + 34 + 4$

$9(X^2 - 4X + 4) + 4(Y^2 - 2Y + 1) = 36$

Factor as Perfect Square Trinomial and divide the whole equation by 36:

$9(X - 2)^2/36 + 4(Y - 1)^2/36 = 36/36$

$(X - 2)^2/4 + (Y - 1)^2/9 = 1$

Center is at (2, 1)

a = 3 and b = 2

$c = SQRT(9 - 4)$

$c = SQRT(5)$

The major axis of the ellipse is parallel to the Y-axis

Vertices are at (2, 4) and (2, -3)

Foci are at (2, 1 + SQRT(5)) and (2, 1 - SQRT(5)).

SOURCE:

COLLEGE ALGEBRA

By

Rees

Sparks

Rees

.

XXI. SOLVING ARITHMETIC SEQUENCES

The following examples are problems involving arithmetic sequences. I included here several sample problems with their solutions.

Problem Number One:

If the first three terms of an arithmetic sequence are 2, 6 and 10, find the 40th term.

To solve the problem we use this formula for finding the nth term of an arithmetic sequence.

$A_n = A + (n - 1) d$

Where, A_n = is the nth term, in the case of our problem it is the 40th term

A = the first term of the sequence , in our problem it is 2.

n = number of terms, in our problem it is 40.

d = the interval of the terms, or the difference of the next term from the previous term, To get d; d = 6 - 2 = 4.

Now, it is time to substitute the values to the formula for solving nth term where the 40th term is to be solved.

$A_n = 2 + (40 - 1) 4$

$A_n = 2 + (39) 4$

$A_n = 2 + 156$

$A_n = 158$.

The 40th term of the arithmetic sequence is 158.

Problem Number Two:

If the first term of an arithmetic sequence is -3 and the eighth term is 11, find d and write the first 10 terms of the sequence.

In this problem,

$A = -3 \quad n = 8 \quad A_8 = 11$

If these values are substituted in the formula for A_n, we have

$11 = -3 + (8 - 1) d$

$11 = -3 + 7d$

$14 = 7d$

$d = 2$

The first ten terms are -3, -1, 1, 3, 5, 7, 9, 11, 13, 15

SUM OF AN ARITHMETIC SEQUENCE

The sum of the first n terms of an arithmetic sequence with first term A and nth term A_n is;

$S_n = n/2 (A + A_n)$ or this formula maybe rewritten as

$S_n = n\{(A + A_n)/2\}$

It can be remembered easily in this form as: "the number of terms multiplied by the mean value or average of the first and last terms."

For an arithmetic sequence with the first term A and common difference d, the sum of the first n terms is;

$S_n = n/2 \{ 2a + (n - 1)d \}$

Problem Number Three:

Find the sum of all the odd integers from 1 to 1111, inclusive.

Solution:

Since the odd integers 1, 3, 5, etc, taken in order from the arithmetic sequence with d = 2, we can first find n from the formula for the nth term;

- $1111 = 1 + (n - 1)2$

$1111 = 2n - 1$

$1112 = 2n$

$n = 556$

$S = 556/2 (1 + 1111)$

$= 278 (1112)$

$= 309,136$

Problem Number Four:

If A = 4, n = 10, A10 = 49; find d and Sn.

Substituting the given values for A, n, and An in the formula:

An = A + (n - 1) d, we get

49 = 4 + (10 - 1) d

49 = 4 + 9d

45 = 9d

d = 5

By using Sn = n {(A + An)/2}, we have

S10 = 10 {(4 + 49)/2} = 5 * 53 = 265

Source:

COLLEGE ALGEBRA (tenth edition) by:

Paul K. Rees

Fred W. Sparks

Charles Sparks Rees

XXII. SOLVING GEOMETRIC SEQUENCES

Arithmetic sequences are formed by addition, whereas geometric sequences are formed by multiplication. Geometric sequences are also called geometric progressions .

A geometric sequence is one in which each term is multiplied by the same number to get the next term. This number is known as the common ratio r,

where r * A_n = A_{n+1} for n = 1, 2, 3, ...

r maybe positive or negative.

Problem One:

Verify whether each of the following sequences is actually a geometric sequence.

(a) 5,000 20,000 80,000 32,0000

Here A = 5000 , r = 20,000/5,000 = 4

(b) 10, 000 5,000 2,500 1,250

A = 10,000 r = 5,000/10,000 = ½

(c) 1,000 2,200 4,840 10,648

A = 1000 r = 2,200/1,000 = 2.2

All of the above are examples of geometric sequences.

The formula for finding the nth term of a certain geometric progression is given as:

A_n = A r ^(n-1)

Where A = first term

r = common ratio

r = A_2/A_1 = A_3/A_2 = A_{n+1}/A_n

n = number of terms

An = the nth term

Problem Number Two:

Find the eighth term of the geometric sequence which begins with ¾ and 3/5.

Solution:

The ratio is: r = 3/5 ÷ ¾ = 4/5

A = ¾ n = 8 , n - 1 = 8 - 1 = 7

Substituting to the formula above :

A 8 = ¾ * (4/5) ^7

= ¾ * 16,384 / 78,125

= [(4,096) *3] / 78, 125

= 12,288/78,125

The formula for finding the sum (Sn) of the first n terms of a geometric sequence with first term A and common ratio r, where r should not be equal to 1 is given as :

Sn = [A (r ^n - 1)]/ r - 1

Problem Number Three:

Find the sum of the first ten terms of the geometric series starting with -5 and 15.

Solution:

r = 15/-5 = -3

A = -5

n = 10

Sn = [-5 (-3 ^10 - 1)] / -3-1

= [-5 (59,049 - 1)] /-4 =-5(59,048)/-4 = -295,240/-4 = 73,810

Alternative formula for Sn:

Sn = (A - r An) / 1 - r

Problem Number Four:

The first term of a geometric sequence is 5 and the fourth term is -320.

Find the eighth term and the sum of the first eight terms.

Solution:

We are given with A = 5, if we first use n = 4 in the formula

An = A r^(n - 1) we obtain,

-320 = 5 r ^3

r ^3 = - 320/5 = -64

r = - 4

We next use n = 8 in the formula for An and Sn

A 8 = 5 (-4)^ 7 = 5 (-16, 384) = -81, 920

S 8 = (A - r An)/ 1 - r

$= [5 - (-4)(-81,920)] / 1 - (-4)$

$= [5 - 327, 680] / 5$

$= -327, 675 / 5$

$= -65, 535$

Problem Number Five:

Find r and A if $S_5 = 1,563$ and $A_5 = 1,875$

Solution:

First, we use the formula for An:

$1,875 = A r^4$ let this be equation (1).

Then we use the formula for Sn,

$1,563 = (A - 1,875 r) / 1 - r$ let this be equation (2)

Solving the second equation for A we obtain,

$(1 - r)(1,563) = A - 1,875 r$

$1,563 - 1,563r = A - 1,875 r$

$312 r + 1,563 = A$ or $A = 312 r + 1,563$, let this be equation (3)

Substituting this value in the first equation, we now have

$1,875 = (312 r + 1,563) r^4$

$1,875 = 312 r^5 + 1,563 r^4$ Or $312 r^5 + 1,563 r^4 - 1,875 = 0$

By using the theorem on rational zeros of polynomial function, we find one of the

Solution to be r = -5.

Substituting r = -5 in the equation (3)

A = 1, 563 + (312) (-5)

A = - 1,560 + 1, 563 = 3

SOURCE: COLLEGE ALGEBRA

Paul K. Rees

Fred W. Sparks

Charles Sparks Rees

XXIII. SOLVING VARIATION PROBLEMS

DIRECT VARIATION

It is a special function which can be expressed as the equation $y = kx$ where k is a constant. The equation $y = kx$ is read "y varies directly as x" or "y is proportional to x. The constant k is called the the constant of variation or constant of proportionality.

Illustration Number One:

The circumference (C) of a circle varies directly as the diameter (d). The direct variation is written as $C = \Pi d$. The constant of variation is Π .

Illustration Number Two:

A teacher makes $10 per hour. The total wage of the teacher is directly proportional to the number of hours (h) worked. The equation of variation is W = 10h. The constant of proportionality is 10.

A direct variation equation can also be written in the form y = kx^n, where n is a positive number. For example the equation y = kx^2 is read " y varies directly as the square of x..

Illustration Number Three:

The area (A) of a circle varies directly as the square of a radius (r) of the circle.

The direct variation equation is A = ∏ r ^2. The constant of variation is ∏.

Sample Problem Number One

Given that V varies directly as r and that V = 50 when r = 5 , find the constant of variation and the equation of variation.

First, write the basic direct variation equation: V = kr

Then, replace V and r by the given values: 50 = k * 5

Solve for k : (1/5) 50 = 5k (1/5 ==è k = 10

Write the direct variation equation by substituting the values of k into the basic direct variation equation: V = 10r

Sample Problem Number Two:

The tension (T) in a spring varies directly as the distance (x) it is stretched.

If T = 20 lbs. when x = 5 inches. Find T when x = 10 inches.

Write the basic direct variation equation: T = kx

Replace T and x by the given value then solve for k:

20 = 5 * k === (1/5) 20 = 5k (1/5) === k = 4

Write the direct variation equation by substituting the value of k into the basic direct

Variation equation: T = 4x

To find T when x = 10 inches. Substitute 10 for x in the equation to solve for T.

T = 4 (10) = 40 lbs.

INVERSE VARIATION

It is a function which can be expressed as the equation y = k/x where k is a constant. The equation y = k/x is read " y varies inversely as x" or "y is inversely proportional to x ." In general, an inverse variation equation can be written y = k/x^n where n is a positive number .

Illustration Number Four:

The equation y = k/x^2 is read "y varies inversely as the square of x."

Given that P varies inversely as the square of x and that P = 10 when x = 2,

Find the variation constant and the equation of variation :.

Set the inverse variation equation : P = k/x^2

Substitute the given values to corresponding variables in the equation and solve for k :

$10 = k/2^2 == (4) \ 10 = k/4 \ (4) == k = 40$

The constant of variation is 40. The inverse variation equation is $P = 40/x^2$

Sample Problem Number Three:

The length (L) of a rectangle with fixed area is inversely proportional to the width.

If L = 10 W = 4, find the length when w = 7.

Write inverse variation equation: $L = k/W$

Substitute the given values to the equation and solve for k :

$10 = k/4 == k = 40$

L = 40/7 or L = 5 and 5/7

JOINT VARIATION

It is a variation wherein a variable varies directly as the product of two or more

other variables. A joint variation can be expressed as the equation Z = kXY, where K is a constant . The equation is read as "Z varies jointly as X and Y.

Illustration Number Five:

The area (A) of a triangle varies jointly as the base and the height. The joint variation equation is written as A = ½ bh. The constant of variation is ½.

COMBINED VARIATION

It is a variation wherein two or more types of variation occurs at the same time.

For example in Physics, the volume (V) of a gas varies directly as the temperature (T) and inversely as the pressure (P). The combind variation equation is written as

$V = k T/ P$

Sample Problem Number Four:

The pressure P of a gas varies directly as the temperature T and inversely as the volume (V). When T = 50 degrees and V = 200 in^3 P = 30lb/in. Find the pressure of a gas when T = 70 degrees and V = 300 in^3.

Write first the basic combined variation equation : $P = kT/V$

Replace the variables by the given values then solve for k :

30 = k (50)/ 200

30 * 200 = 50*k

(1/50) 6000 = 50 k (1/50)

k = 6000/50 = 120

P = 120 T/V =è P = 120 (70)/300

(1/300) 300 P = 8400 (1/300)

P = 8400/300 = 28 lb/ in.

SOURCE : INTERMEDIATE ALGEBRA

AN APPLIED APPROACH

By Aufmann/

XXIV . Complex Numbers: Power of I (Additional)

We did not define nth root of a certain number as if n is even and a is negative. For instance square root of -9 was not defined . In particular we did not define SQRT(a) if a is a negative since there is no real number whose square root is negative. Thus we cannot solve an equation such as X^2= -81 using only real numbers. We may extend the real number system to a larger system called the complex numbers system. To do this we first define the imaginary number i :

i^2 = - 1 or i = SQRT(-1)

Therefore -i = i^3

Since i^3 = (i^2)(i) = (-1) (i) = -i

i^4 =1

Since i^4 = (i^2) (i^2) = (-1)(-1) = 1

$i^5 = i$

Since $i^5 = (i^4)(i) = (1)(i) = i$

Sample Exercises

Number One: Show that $i^{27} = -i$

$i^{27} = (i^{24})(i^2)(i)$

Since $i^{24} = (i^4)^6 = 1^6 = 1$ and $i^2 = -1$

Therefore $i^{27} = (1)(-1)(i) = -i$

Number Two: Find i^{105}

$i^{105} = (i^{104})(i)$

Since $i^{104} = (i^4)^{26} = 1^{26} = 1$

Threfore $i^{105} = (1)(i) = i$

Number Three: Find i^{307}

$i^{307} = (i^{304})(i^2)(i)$

Since $i^{304} = (i^4)^{76} = 1^{76} = 1$ and $i^2 = -1$

Therefore $i^{307} = (1)(-1)(i) = -i$

Number Four: Find i^{1002}

$i^{1002} = (i^{1000})(i^2)$

Since $i^{1000} = (i^4)^{250} = 1^{250} = 1$ and $i^2 = -1$

Therefore $i^{1002} = (1)(-1) = -1$

SOURCE:

COLLEGE ALGEBRA BY

REES

SPARKS

REES